# SELL YOUR RESEARCH

Alexia Youknovsky · James Bowers

# SELL YOUR RESEARCH

## Public Speaking for Scientists

 Springer

Alexia Youknovsky
Agent Majeur
Paris, France

James Bowers
Agent Majeur
Paris, France

ISBN 978-3-030-34180-0      ISBN 978-3-030-34181-7   (eBook)
https://doi.org/10.1007/978-3-030-34181-7

The SELL Method™ is a registered trademark (n° 4616654), property of Agent Majeur.

This Springer imprint is published by the registered company Springer Nature Switzerland AG
The registered company address is: Gewerbestrasse 11, 6330 Cham, Switzerland

*Every scientist can become a great speaker.*
*It is simply a matter of hard work and technique.*
*Here is the method.*
*The rest is down to you…*

# The Authors

**Alexia Youknovsky** *Founder and CEO, Agent Majeur*
A chemical engineer by training, with experience in both research and consulting, Alexia was also an actress for six years. Twelve years ago, she created Agent Majeur, a science communication agency to provide specific services for stakeholders in research and innovation. She related to the constant pressure on scientists to communicate their activities and felt her combination of skills were a valuable asset. Now, she leads a team of science communication professionals with the objective of helping researchers promote their work through training, strategic planning, event production, graphic design, and various other activities.

**James Bowers, Ph.D.** *Science Communication Consultant and Coach.*
Passionate about science and media, British-born James obtained his Ph.D. in Molecular Biology and Physiology in 2014. Upon completion, he was quickly swept up with the wave of young scientists choosing to devote themselves to research dissemination over the lab life. After the experience of presenting science to the general public in museums and events, he earned an M.Sc. in Science Media Production from Imperial College London. He then worked in factual TV production and explained science on several shows. At Agent Majeur for over two years, James trains in public speaking and science writing, consults on a variety of science communication projects and hosts events.

# About Us

Research and innovation is about new discoveries and novel ideas, but at Agent Majeur we believe it can be even more than that. By pushing research and scientific breakthroughs into the spotlight, we can help them grow and be noticed.

Our slogan represents our vision: "Adding value to science." We help scientists and companies to promote their work in the most appropriate way to suit their need: events, presentations, communication campaigns, showrooms, brochures and more. To succeed, one requires an understanding of the research landscape and the capacity to grasp complex topics. Hence our team of communicators are themselves educated to engineer or Ph.D. level.

During our training courses, we provide tools to overcome modern challenges in science communication. Whether it be promoting a research project, pitching an innovation, drafting a scientific article, or enhancing funding applications, we train in written and spoken communication of science. In public speaking alone, we have coached over 2,500 people, offering more than 15 different training programmes.

There is a distinct lack of handbooks or guides outlining steps for preparing a presentation. This book presents the SELL Method, which we have used as the basis of our training courses since 2008. We know it works, so feel free to use it too!

# Get in Touch

We would love to open a dialogue with you. It is a great way for us to stay on track and challenge our practises. What services could we create to meet your needs? What training courses do you feel are missing? What topics would you like to see addressed in our blog?

**You can contact us here**: contact@agentmajeur.fr

**For more science communication advice**: www.agentmajeur.com/blog

# Contents

**1 Introduction**     1

**2 The SELL Method: Prepare Your Presentation**     5
2.1   What Is Communication?     6
    2.1.1   Conversation     7
    2.1.2   Communication     7
    2.1.3   The Communication Model     8
2.2   The SELL Method     11
2.3   Step I: Skeleton     12
2.4   Step II: Envelope     13
2.5   Step III: Life and Logistics     13
2.6   Let's Get on with It!     14

**Part I   First Step: Build Your Skeleton**

**3 Know Your Context: Types of Presentation**     17
3.1   Audience     18
    3.1.1   Experts     18
    3.1.2   Students     19
    3.1.3   General Public     19
    3.1.4   Young People     19
    3.1.5   Journalists     20

|  | 3.1.6 | Managers | 20 |
|  | 3.1.7 | Funders | 21 |
| 3.2 | Format |  | 21 |
|  | 3.2.1 | Audience Size | 21 |
|  | 3.2.2 | Interaction | 22 |
|  | 3.2.3 | Duration | 23 |
|  | 3.2.4 | Your Stage | 25 |
| 3.3 | Occasion |  | 26 |
|  | 3.3.1 | Meeting | 26 |
|  | 3.3.2 | Conference | 27 |
|  | 3.3.3 | Site Tour | 27 |
|  | 3.3.4 | Poster Session | 27 |
|  | 3.3.5 | Thesis Defence | 28 |
|  | 3.3.6 | Prize Ceremony | 28 |
|  | 3.3.7 | Competition | 28 |
|  | 3.3.8 | Camera | 29 |
|  | 3.3.9 | Interview | 29 |
| 3.4 | Logistics |  | 29 |
|  | 3.4.1 | What? | 30 |
|  | 3.4.2 | Who? | 30 |
|  | 3.4.3 | Where? | 30 |
|  | 3.4.4 | When? | 30 |
|  | 3.4.5 | How? | 31 |
|  | 3.4.6 | Why? | 31 |
|  | 3.4.7 | How Many? How Much? | 31 |
| 3.5 | Moving on… |  | 31 |
| **4** | **Define Your Message** |  | **33** |
| 4.1 | Four Questions to Define Your Message |  | 34 |
|  | 4.1.1 | What Is Your Objective? | 35 |
|  | 4.1.2 | What Are Your Audiences' Expectations? | 35 |
|  | 4.1.3 | What Arguments Do You Have to Achieve Your Objective with This Specific Audience? | 36 |
|  | 4.1.4 | What Is Your Message? | 36 |
|  | 4.1.5 | Putting It into Action | 37 |
| 4.2 | Case Study I: The Headteacher |  | 37 |
|  | 4.2.1 | Science Talks | 38 |

| 4.3 | Case Study II: The Bionics Professor | 40 |
| | 4.3.1 | Experts | 40 |
| | 4.3.2 | General Public | 40 |
| | 4.3.3 | Different Messages | 43 |
| 4.4 | Case Study III: The Entrepreneur | 43 |
| | 4.4.1 | Investors | 43 |
| | 4.4.2 | Coffee Shop Managers | 44 |
| | 4.4.3 | Clients | 44 |
| 4.5 | Messages for Mixed Audiences | 44 |
| | 4.5.1 | Prioritise Your Audience | 48 |
| | 4.5.2 | Find the Common Link | 48 |
| 4.6 | Your Turn… | 48 |

| **5** | **Establish Your Plan** | 51 |
| 5.1 | A Plan: The Basics | 52 |
| | 5.1.1 | What is a Plan? | 52 |
| | 5.1.2 | Examples of Classical Structures | 52 |
| | 5.1.3 | When to Give Background Knowledge | 53 |
| | 5.1.4 | What Does a Plan Look Like? | 53 |
| 5.2 | Mind Maps | 54 |
| | 5.2.1 | How to Use a Mind Map | 54 |
| | 5.2.2 | Benefits of Mind Mapping | 56 |
| | 5.2.3 | Putting It into Action | 56 |
| 5.3 | Case Study IV: The Technological Innovation | 57 |
| 5.4 | Case Study V: The Success Story | 59 |
| 5.5 | Case Study VI: The Controversial Topic | 60 |
| 5.6 | Before We Move on… | 62 |

**Part II   Second Step: Make Your Envelope**

| **6** | **Hook Your Audience** | 65 |
| 6.1 | Why Hook Your Listeners? | 66 |
| 6.2 | When to Hook Your Listeners | 66 |
| | 6.2.1 | Introduction | 67 |
| | 6.2.2 | Conclusion | 67 |
| 6.3 | Hook, Line and Sinker | 68 |
| | 6.3.1 | Quote | 68 |
| | 6.3.2 | Joke | 69 |
| | 6.3.3 | Anecdote | 70 |
| | 6.3.4 | Analogy | 71 |

| | | | |
|---|---|---|---|
| | 6.3.5 | Question or Challenge | 72 |
| | 6.3.6 | Picture, Drawing or Photo | 73 |
| | 6.3.7 | Video | 73 |
| | 6.3.8 | Piece of Music | 74 |
| | 6.3.9 | Link with Current or Historical Facts | 75 |
| | 6.3.10 | Sample | 75 |
| | 6.3.11 | Demonstration | 76 |
| 6.4 | | Knitting Your Hook into Your Talk | 76 |
| 6.5 | | Be Comfortable with Your Hook… | 77 |
| **7** | | **Design Your Slides** | **79** |
| 7.1 | | What's the (Power) Point? | 80 |
| 7.2 | | A Point with Power | 80 |
| | 7.2.1 | Rules Are Meant to Be Broken | 81 |
| | 7.2.2 | Keep It Slick | 81 |
| | 7.2.3 | Use Readable Text | 82 |
| | 7.2.4 | Be Visual | 82 |
| | 7.2.5 | Be Coherent | 83 |
| | 7.2.6 | Simplify Your Graphs | 83 |
| | 7.2.7 | Use Animations | 83 |
| | 7.2.8 | Think About Your Transitions | 84 |
| | 7.2.9 | Point Out Important Figures | 84 |
| | 7.2.10 | Lighten Your Content | 85 |
| | 7.2.11 | Choose Quality Images | 85 |
| 7.3 | | Graphic Design Rules | 85 |
| | 7.3.1 | Contrast | 86 |
| | 7.3.2 | Alignment | 86 |
| | 7.3.3 | Proximity | 86 |
| | 7.3.4 | Repetition | 87 |
| 7.4 | | Technical Glitches | 87 |
| 7.5 | | Slide Improvements | 88 |
| | 7.5.1 | Initial Thoughts | 89 |
| | 7.5.2 | Give It Space, Let It Breathe | 89 |
| | 7.5.3 | Don't Give Everything Away | 89 |
| | 7.5.4 | Help Your Data Speak for Themselves | 90 |
| | 7.5.5 | Point Out the Comparisons | 92 |
| 7.6 | | In Conclusion… | 93 |
| **8** | | **Create Your (Additional) Supporting Materials** | **95** |
| 8.1 | | What Tools Do You Have? | 96 |
| 8.2 | | Boards | 96 |

|  | 8.2.1 | Erasable Boards | 97 |
|  | 8.2.2 | Non-erasable Boards | 97 |
| 8.3 | Objects | | 98 |
|  | 8.3.1 | Scientific | 98 |
|  | 8.3.2 | Props | 99 |
|  | 8.3.3 | 3D Metaphors | 99 |
| 8.4 | Demonstrations | | 100 |
| 8.5 | Multimedia: Videos and Sound | | 101 |
|  | 8.5.1 | Videos | 101 |
|  | 8.5.2 | Animations | 102 |
|  | 8.5.3 | Sounds | 102 |
| 8.6 | Before We Move on… | | 103 |

**9    Popularise Your Science**          107
| 9.1 | What Is Science Popularisation? | | 108 |
|  | 9.1.1 | When to Popularise | 108 |
|  | 9.1.2 | Are You Making Yourself Clear? | 110 |
|  | 9.1.3 | Signpost Your Talk | 111 |
|  | 9.1.4 | How to Popularise | 111 |
| 9.2 | Make a Connection | | 111 |
|  | 9.2.1 | Technical Impact | 112 |
|  | 9.2.2 | Personal Insight | 112 |
|  | 9.2.3 | Funny Tales | 112 |
| 9.3 | Give Context | | 113 |
|  | 9.3.1 | Choose a Character | 113 |
|  | 9.3.2 | Make It Relatable | 113 |
| 9.4 | Be Concrete | | 114 |
|  | 9.4.1 | Jargon | 114 |
|  | 9.4.2 | Data | 115 |
|  | 9.4.3 | Visuals | 116 |
| 9.5 | Try It Out… | | 116 |

**Part III    Third Step: Breathe Life and Fine Tune Logistics**

**10    Master Your Non-verbal Communication**          119
| 10.1 | What is Non-verbal Communication? | | 121 |
| 10.2 | Voice | | 122 |
|  | 10.2.1 | Breathing | 122 |
|  | 10.2.2 | Speed | 122 |
|  | 10.2.3 | Tone | 123 |

|  | 10.2.4 | Volume | 123 |
|  | 10.2.5 | Diction | 124 |
| 10.3 | Body Language | | 124 |
|  | 10.3.1 | Posture | 125 |
|  | 10.3.2 | Gestures | 125 |
|  | 10.3.3 | Facial Expressions | 126 |
|  | 10.3.4 | Movements | 127 |
| 10.4 | Eye Contact | | 127 |
| 10.5 | Before We Move on… | | 128 |

**11   Manage Your Anxiety**                                    129
11.1   Practise, Practise, Practise                             130
       11.1.1   To Learn or Not to Learn                        130
11.2   Enter the Stage                                          132
11.3   Breathe                                                  132
11.4   Project Yourself                                         133
11.5   Respect Your Body                                        134
11.6   Before We Move on…                                       135

**12   Perfect Your Performance**                               137
12.1   Microphone Choice                                        138
       12.1.1   Gooseneck Microphone                            139
       12.1.2   Handheld Microphone                             139
       12.1.3   Headset                                         140
       12.1.4   Clip-on Microphone                              140
12.2   Chair(wo)Man                                             140
12.3   Technical Rehearsal                                      141
       12.3.1   Technical Rehearsal Checklist                   142
12.4   Lights, Camera…                                          142
12.5   Final Touches                                            143
12.6   That's It…                                               143

**13   Final Pep Talk**                                         145

**14   The Story Behind the SELL Method**                       147

**Speaker References**                                          149

**Bibliography**                                                153

# 1

# Introduction

Scientists present their work all the time.

Whether you are a researcher, engineer, Ph.D. student, post doc, innovator or technician, you will definitely be brought to speak about your research in public at some point or another. Many people do genuinely enjoy presenting their work. If you are one of those, keep on reading! In this book, you will find tons of valuable ideas to help you get better in your preparations. If that doesn't sound like you, then we urge you to continue reading too. This book is also aimed at scientists like yourself who are asked to give a talk but have no clue where to start.

We know that as a scientist you have a never-ending to do list and preparing a presentation is likely never right at the top of it. Maybe you even refuse out right to put in too much effort because that precious time could be better spent doing other things. We get it. You would rather work on that hefty grant proposal, help a Ph.D. student make it through their viva, complete a stubborn experiment or finally finish that paper that you have been drafting for three years now. But as you already know, talking about your research is unavoidable. More than that, there are a heap of extraordinary benefits to giving a memorable talk: you will feel good about yourself; your research will make more of an impact; you could find more funding; you may attract new collaborators, students or staff; and the list goes on.

We bet that, with our help, you will actually enjoy speaking in public. Yes, it really can be a pleasant experience—we promise!

We work with scientists on anything from one-minute pitches to hour-long plenary talks and we assure you that anyone can become a successful public speaker. The payoff from knowing how to talk about your work is

© Springer Nature Switzerland AG 2020
A. Youknovsky and J. Bowers, *SELL YOUR RESEARCH*,
https://doi.org/10.1007/978-3-030-34181-7_1

immeasurable. Once you have prepared a good presentation and received the positive feedback you deserve, you will feel more confident the next time. The more you practise and put your mind to it, the better you will get.

In France, one of our favourite desserts is the chocolate soufflé. Sweet tooth or not, this spongy pudding with its soft centre and crunchy top is a delight. If you have ever tried to bake one, then you will know how notoriously difficult it is to achieve the distinctive fluffy texture. Even experienced chefs can fudge it because the conditions required for the middle to rise in just the right way are extremely specific. Some would say it is the artistry of baking at its best.

Nonetheless, people of course can (and do) learn how to make the perfect chocolate soufflé. As intricate a process as it may be, to succeed the following is needed:

- specific ingredients;
- a killer recipe;
- the best utensils;
- a dash of personal flair.

We believe that preparing a science talk is the same. As scientists, researchers or engineers, you already have all the *ingredients* you could possibly need—the content. You are the expert. You hold or have access to the extensive knowledge necessary to present your topic. But which bits should you use in your next presentation? In this book, not only do we explain how to select these ingredients, but we also provide you with the rest: a killer *recipe,* tips on how to pick the best *utensils* and techniques to help you add your own *personal flair.*

The biggest challenge in science communication is accepting that most of what you know about your subject, you won't have time to mention. And that is absolutely fine. Your speech should be a window into an aspect of your work, not a recital of every reference in your Ph.D. thesis. In the following pages, you will find everything you need to know to prepare and deliver your presentation in the most efficient way possible. We offer you all the right questions you should ask yourself, in the most logical order.

This book is here to help you make the most of your public speaking opportunities, whether it be a talk, pitch, or seminar. We provide you with a bank of examples from our vast experience in the hope of inspiring you to come up with your own. Our recipe will help you stay on the right track to SELL YOUR RESEARCH.

Let's face it, you will be talking about a subject that you know, love and dedicate a huge chunk of your life to. You can enjoy it. And your audience deserves to have a great time too!

# 2

# The SELL Method: Prepare Your Presentation
## Our Killer Recipe for Concocting a Talk

We have all been awestricken by the capacities of an athlete competing in the Olympic Games at one time or another. Just think of the sheer skill required for a pole jumper to scale a bar six metres high with room to spare. It is easy to feel like their abilities are completely unattainable to us mere mortals. Upon seeing them in action, we often forget that athletes all started somewhere. Gymnasts must spend a lifetime honing the flexibility in their limbs. Runners dedicate their lives to keeping their body in shape so that they can continue to push themselves further and faster.

The one thing all good athletes have in common is lots of training, in other words: preparation. Yet, sportsmen and women cannot be sure they will win, all they can do is to train as best they can to be at the top of their game on the day of their competition. Throughout this book we talk a huge amount about preparation. Every single chapter discusses ways of preparing yourself and we will talk about it until we are blue in the face. There is a good reason for this: when you do your presentation, talk or seminar, you only get one chance to get it right. Like an athlete, to be the best you can be, the only thing you can do is to train yourself in every way possible.

All scientists can do a presentation, but not all know how to prepare. In this book, our objective is to hand you the materials and exercises you need to train yourself, so that you can become a brilliant speaker, capable of delivering Olympic-grade science presentations.

There is a common misconception that if you are not instinctively a good communicator, then you can't learn how to be one. That's wrong. Even talented speakers must put effort into preparing themselves. Whilst it may seem paradoxical, the more you prepare and rehearse, the more natural your talk

A. Youknovsky and J. Bowers, *SELL YOUR RESEARCH*,
https://doi.org/10.1007/978-3-030-34181-7_2

will sound. Successful public speaking appears effortless to the audience, which can actually be quite frustrating because it means that all the hard work effectively becomes invisible. Bear that in mind the next time you see a good presentation. If someone looks calm and cool whilst they are giving a speech, it is usually because they are well prepared rather than having some superhuman talent for public speaking or relying on natural born confidence.

Now, imagine you are sitting at a piano. You press the keys in front of you and each one makes a different sound. Having never learnt to play before, you hit the keys in any random order and, although a sound comes out, you are not able to play music *per se*. Even someone with raw musical talent will never play a piece of Mozart without first learning how. Lessons and/or a lot of practice are necessary to acquire the appropriate skills. Only then can a truly talented person tap into their gift.

We often notice this in our training courses. We can see when speakers with potential don't put the work into improving, because it shows. On the other side of the coin, those who are perhaps less naturally skilled in presenting are more than capable of giving some of the best talks, if they put in the effort required.

We believe that everyone has a gift for speaking that they can hone. Part of the skill is making people feel what we call in French, a *coup de cœur*. A special sensation that you—and *only* you—can offer the world. It could be your great sense of humour; loveable character; endearing vulnerability; or infectious passion for a precise South American species of toad. Everyone has that *je ne sais quoi*. Let's find yours!

In this chapter, we provide some theory about communication and what it can tell us about delivering a memorable presentation. We will then apply that knowledge to our unique technique, which forms the basis of this entire book: the SELL Method. Each chapter that follows will take you through a step in this fundamental process.

## 2.1    What Is Communication?

Before you begin, it can help to know what we are actually referring to by the term 'communication'. Here we provide the basic principles in the form of a communication model, for you to keep in mind along the way when preparing and delivering your presentations.

## 2.1.1 Conversation

An essential fact about communication is that it differs from conversation. Imagine yourself chatting to your mate about something that interests you, say the latest episode of your current Netflix binge. During this interaction, you and another person will both share and learn information from one another. It's an exchange: speech is essentially bilateral.

Conversation has this give-and-take nature and, as such, when you are speaking to someone, you are improvising. You don't prepare what you are going to say in advance. If someone asks you a question in a conversation, you formulate an answer on the spot. Likewise, when someone tells you a joke that you find funny, you may laugh spontaneously. If someone explains an idea to you, you can immediately signal back and make it clear whether you agree or disagree with the statement.

This improvised interaction is not limited to speech. As a human being, when you are participating in conversation, your body language is spontaneous too. When you feel genuinely engaged with someone, you will mirror their posture. You may fold your arms if you feel uncomfortable or look away if you are being dishonest. Most of the time, you aren't aware of much of these indicators, but they are definitely there.

The next time you are having a conversation, try analysing the situation. Pay attention to yours and your partner's movements, focus on the fact that you are talking and listening, responding and reacting. You will probably find that it immediately feels uncomfortable, as you lose spontaneity.

The natural flow of improvised conversation is an important human ability and this in-the-moment behaviour can be very enriching when forming new ideas. It makes sense, right? We are social beings, after all. Talking is what we do.

## 2.1.2 Communication

Contrary to conversation, communication is *not* spontaneous. When you communicate, you prepare in advance with an objective in mind. The reason for communicating in the first place is whatever you, as a speaker, want people to do or think as a result of your presentation. This could be:

- "I want to convince you that…"
- "I want you to fund…"
- "I want you to think that…"
- "I want you to understand…"

- "I want you to buy…"
- "I want you to vote for…"
- and many, many more!

As a communicator, you effectively become a transmitter. Rather than a shared interaction with input from either side, as with conversation, communication is the transfer of information from one side to the other, with a specific purpose. Simply put, communication is a global term meaning transmission of an idea that will help you achieve your goal. This idea, that we call your *message*, is sometimes also referred to as the through line. In communication, you must take the time to define it before taking to the stage. Once you have established a clear message, it is easier to build a strong talk around it.

### 2.1.3    The Communication Model

Researchers have racked their brains for over 70 years thinking about how to define communication. Many of them have attempted to identify the components involved and represent them in a schematic way. The first of these was proposed by Shannon and Weaver, in 1949. For many, it is considered the reference model. Other authors since then have adapted this schema according to their own interpretations but nonetheless, each of the models essentially contains the same basic elements.

Without going into a long-winded description of these models, we propose a simplified version, summarised in the following communication model (Fig. 2.1). In public speaking, you are the sender and each person in your audience is a receiver. The message is the content that you want your receiver to retain from your talk.

In an ideal world, when you express yourself, there will be no difference between what you want to say and how your spectators interpret that information. You would hope that when your audience listens to you, they understand you. But unfortunately, the reality is much more complex. Transmission of your message can be modified and thus distorted by what we call filters. These are barriers that we all have, which can influence how we understand and react to a piece of communication. In a successful presentation, the speaker has done enough work to make these filters disappear.

Below, we define the various components of the communication model.

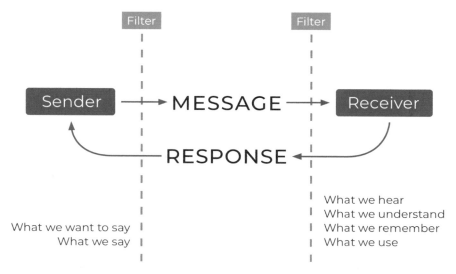

**Fig. 2.1** The communication model. ©Agent Majeur

### 2.1.3.1 Sender

In communication, the sender is a transmitter of information or, in other words, a speaker or presenter. The purpose of this book is to make you a good sender. To start with, you should have a clear idea of who your receivers are, what they expect from your presentation and what filters they may have. By considering these parameters, you can develop a talk that suits your audience in the most appropriate way and ensures the successful transmission of your message.

### 2.1.3.2 Receiver

The receiver is your listener or spectator—your audience. They can be fellow scientists, members of the public, funders, students and any other category of specialists or lay persons. The only common denominator of your potential receivers is that they will all be human. That makes your list of possible audience expectations pretty much infinite!

For your presentation to function, you need to make sure that your receivers *receive* your message. If it is successfully transmitted, your audience will provide you with the response you were hoping for: laughter, applause, their confidence in your project, a request for a collaboration, or a promise of funding, for example.

### 2.1.3.3   Message

The message of your talk is the one main point that you want your audience to remember when your communication is over. Once you have defined your message, your talk should drill that idea into your audiences' heads as effectively as possible. Regardless of how you intend to convey this message, you *have* to make it memorable.

Whilst your audience will remember different bits of information from your talk, everybody should leave with the same message at the forefront of their minds. If you achieve this, then your presentation has been a success! This part is so important that it has its own chapter—see Chap. 4.

### 2.1.3.4   Filters

Between you and your audience, there are two filters, or barriers, through which your message needs to pass. As a sender you will have your own filters, such as prior level of knowledge, language used or cultural references. By knowing your own filters, you can ensure that what you say is a true reflection of what you *want* to say by carefully regulating *how* you say it.

The opposite is true for your audience. They will have their own set of filters. The first filter is: are they listening to you? It is your responsibility, as a sender, to make sure your audience can hear you and that they feel intrigued by what you say. Second, it is important that they understand you. Have a think about this example: you may be from the UK and your receivers from the USA. Even though you are both native-English speakers, references to British research funding procedures may not have the same relevance to the people sitting in an American audience.

It is your role, as a sender, to take into account the filters which belong to both you and them. That way, you can vastly improve the delivery of your message. Be aware that, unlike you, your receivers do not always have previous knowledge of your subject. Particularly in science talks, one of the main filters you will come across is your audience's level of technical knowledge. Even experts may not be familiar with jargon which is commonplace to you. Perhaps they do not know the right terminology, or maybe the way you use or understand it is different. We will come back to this in Chap. 9.

### 2.1.3.5 Response

You will also see on the schema that there is a response between receiver and sender. In the case of a presentation is not necessarily referring to a verbal answer as with a conversation. It may simply be an action taken by the receiver: buy a product, ask a question, applaud etc.

Successful talks involve a certain level of strategy, by integrating the knowledge you have about your audience and their filters. Once you have your message, you should build your presentation around it. All the choices that you make serve the purpose of helping you get your point across. You can play around with it by adding humour, character, supporting materials, a relevant catch line and so on. We will cover all these tools and techniques over the course of the chapters in this book.

## 2.2   The SELL Method

Now you understand the theory of communication. To prepare a good science presentation, you need to put these elements into practice in an organised and efficient way. And for this, we've created a killer recipe!

Preparation is everything. You will not make an effective talk without it. Over the past 12 years, Agent Majeur has been training scientists, medical professionals, innovators, technicians, communicators and more to make good presentations. We have grouped the different parts of preparing a presentation into three easy-to-use sections, the SELL Method (Fig. 2.2):

- build a Skeleton,
- make an Envelope,
- breathe Life and fine tune Logistics.

By using this structure, you can work through the stages of preparing your speech. Our method will help you to order your thought process and formulate your talk well, right from the beginning. Be ready to start this process as early as possible before your deadline, because it takes time. You will not necessarily find all the responses to the questions at the beginning. Make, re-make, write, re-write, perfect and refine; these are the steps you will go through. That's why we advise you, where possible, to begin your preparation at least two weeks before the day.

**Fig. 2.2** The SELL Method for preparing a science talk. ©Agent Majeur

## 2.3 Step I: Skeleton

As the name suggests, your skeleton is the basis of your talk. It provides a framework upon which you can build a presentation that is perfectly adapted to your content, your audience and your own personality. By piecing together your skeleton, you lay the groundwork for the rest of your preparation. It adds value to your presentation overall and deserves the time it takes.

When preparing your skeleton, you will consider why you are doing your presentation and under what conditions. You will take the time to establish a clear message by analysing your objectives and identifying the expectations of your audience. Planning is also a vital component of your skeleton. By the end of this step, you will have a solid message and a strong plan for your talk. The skeleton is composed of the following sections:

- Know your **context**
- Define your **message**
- Establish your **plan**.

## 2.4    Step II: Envelope

Once you know your message and how you want to structure your presentation, you can start wrapping it up. You will build a firm body around your skeleton using a range of different tools.

During this section, you will also choose supporting materials best adapted to your needs. Will you use PowerPoint slides or live demonstrations? Do you want to make your presentation interactive? Furthermore, what angle should you take in sharing your message: funny, serious or mysterious? And will you be using anecdotes or examples?

The envelope is composed of the following sections:

- **Hook** your audience
- Design your **slides**
- Create (additional) **supporting materials**
- **Popularise** your science.

## 2.5    Step III: Life and Logistics

When you get to this third stage, you will have constructed the intellectual side of your talk. You now need to be capable of carrying your presentation physically. When making that soufflé we were talking about, the final touch is the personal flare. The life of your presentation is your personality, physical engagement and delivery. These things all contribute to the connection you create with your audience. To make your talk as engaging as possible, you need to breathe life into it.

This section is the hardest to grasp, but once you have found your own personal style, you can use it to kick ass in every talk you ever give. We suggest that you start adding life to your talk by practising it out loud several times. You will also work on making eye contact with your audience, maintaining good posture, mastering your body language and using breathing exercises to stave off anxiety. You should then be able to express yourself in your own personal way and keep people engaged with you as a presenter.

There are certain logistical elements to take into consideration too. Things such as knowing how to use your microphone or making sure you're standing in the light on your stage all contribute to providing a polished performance. We combine life and logistics in our final step because we believe they go hand in hand to making your presentation a success.

Life and Logistics is composed of the following sections:

- Master your **non-verbal communication**
- Manage your **anxiety**
- Perfect your **performance**.

## 2.6     Let's Get on with It!

Now that you know the basics of the SELL Method, we can look into applying it when preparing your presentations and help you to give the best possible talk.

# Part I

## First Step: Build Your Skeleton

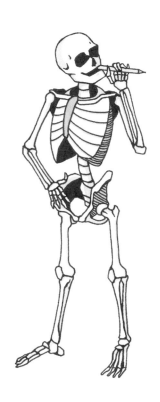

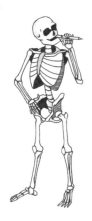

# 3

# Know Your Context: Types of Presentation
## Establish the Setting

Let's imagine for a second that you decide to buy a plot of land upon which to build a house. You know that you want five bedrooms, three bathrooms and a large terrace with a family-size barbeque—we can all dream! However, before you can start designing your new home, you need to know as much as possible about the terrain. You must take stock of the dimensions of the plot, any variations in the topography and what resources you have. These bits of information will influence how you can build the house and provide you with a set of parameters to follow. A science talk is no different.

We repeat, the best way to give a good talk is to be prepared. The best way to be prepared is to ask yourself the right questions early on. At the beginning of your preparation, you need to scope out your plot of land to define what the limitations of your talk will be. In short, you need to know why you are

© Springer Nature Switzerland AG 2020
A. Youknovsky and J. Bowers, *SELL YOUR RESEARCH*,
https://doi.org/10.1007/978-3-030-34181-7_3

giving your talk, what is expected of you and the tools at your disposal. That is why we recommend you ask yourself the following questions:

- Who will be there?
- What is the format?
- What type of presentation is it?

In this chapter we will provide you with an abundance of the most common answers to these questions. Every context is unique and so the combination of responses will differ for each presentation you do.

## 3.1   Audience

There is always a reason as to why you are giving a talk. Your audience will have a motive for coming to see you present your research or technological innovation too. You should know what that driver is. Perhaps they are a fellow expert carrying out research in the same field, vegan cosmetics, for example. Or perhaps they are a group of Instagram influencers who are intrigued to find out more about the research behind your lipstick pigment free of animal products.

When thinking about your audience in this way, you should also be aware in advance of whether they are likely to be critical of your work, as this will help you establish the level of 'convincing' required. You need to take this into account when preparing. What are they interested in? How can you adapt your presentation to suit them? To understand more about how to deal with different audiences, see Chap. 4. But for now, here are some different types of audience that you should consider, along with their expectations.

### 3.1.1   Experts

We can split the experts who could be present at your talk into two categories: specialists in your domain and those who work in another field. The people in these two groups do not have the same level of knowledge about your subject and their expectations can differ. Here is a non-exhaustive list of reasons why experts would come to one of your talks:

- to move their own research forward,
- to find other researchers with whom to collaborate,
- to develop an international research project,
- to grow their own professional network.

## 3.1.2 Students

Having started higher education but not yet working life, students are a very particular type of public. In general, they are curious and have a desire to take something away from the opportunity to attend. They are in a pedagogic mind frame and generally hold a specific interest in the talks they choose to see. Student's expectations include:

- to move forward in their career aspirations,
- to discover careers that are open to them or how they can get hired,
- to find a placement or first job,
- to enhance their knowledge of a topic they are studying,
- to learn more about a subject that interests them.

## 3.1.3 General Public

When your audience consists of members of the public, the people present can have different expectations based on their geographical, social or cultural origins along with their career, studies or hobbies. In this paragraph we are mainly interested in the most frequent motivations of adults for attending your presentation:

- to find out what scientists are working on,
- to understand how your research affects their own lives,
- to reinforce their understanding of a field that interests them,
- to learn new things or educate themselves out of pure curiosity.

## 3.1.4 Young People

Something specific to children is that they have probably not decided to attend your talk of their own accord. It is surely an adult—a parent, guardian or teacher—who has made that choice for them. Another big difference between children and adults is their attention span and level of concentration. Four-year-olds will struggle if you speak for longer than 10 min, whilst 11-year-olds can stay alert for more like 45 min. You need to factor this into a presentation for young people and adapt your talk around their abilities.

You must vary activities regularly and be aware of school curriculums for the age of the young people you are speaking to. Pay attention to keeping them engaged and inspired with new exciting ideas. Young people will he hoping for the following:

- to be entertained,
- to feel inspired for their future,
- to learn more about a subject that interests them,
- to discover fun things.

### 3.1.5 Journalists

It is not every day that journalists will attend one of your talks. Their presence is a great opportunity to get people talking about your work over and above your usual scope. To make them want to write an article about your research or cite you in a paper, you should focus your speech on something brand new or unique, because journalists are permanently looking for that scoop.

Don't forget that, via journalists, you are communicating with their future readers. Depending on whether the journalist is working for a national newspaper or, conversely, a specialised review, the level of simplification you should use will differ. Journalists expect:

- to find an interesting subject for an article/report,
- to seek out scientists to interview for an upcoming article/report,
- to add scientists to their database of contacts,
- to inform themselves of scientific progress in a given field.

### 3.1.6 Managers

Within the framework of your research laboratory or company, at some point you will have to present your work to hierarchical superiors. Whether it is the leader of your institute, head of department or R&D director, their expectations are generally the following:

- to become a national or global reference in a particular field,
- to integrate technological innovations into future products or services to gain a competitive advantage,
- to keep up with the state of the art and be up to date with competitors,
- to use budgets that they control well,
- to ensure the successful development of a project under their responsibility.

### 3.1.7  Funders

Investors or funders can have different expectations based on whether they are public or private. Nevertheless, they share the same basic reasoning: they want to know that their money will be used wisely if they put their trust in you. Other objectives include:

- to identify applicants or teams with the highest potential for success,
- to develop knowledge in a certain domain that is useful to the community,
- to create jobs in a geographical region,
- to generate revenue thanks to a reliable and profitable investment,
- to inform themselves of the latest scientific advances in a field where they normally invest.

## 3.2  Format

Understanding the format of your presentation will give you a framework within which you can build your talk. For example, there is no point preparing a 20-min PowerPoint presentation when the organiser has only booked you in for a 5-min slot with no access to any projector equipment. You should be told these constraints when you are invited to talk, but if not, make sure you ask.

The format is essentially the perimeter surrounding your presentation: audience size, type of interactions with the audience, duration, space and equipment available. It provides you with a guide of what you can and cannot do. We consider format to be the conditions which are, most often, imposed by the organiser of the event where you are speaking or whoever invited you to talk. Whilst you can request a flipchart for your talk or table to put a sample on, you will commonly have little or no control over many of the parameters in this section.

### 3.2.1  Audience Size

How you prepare your presentation will change depending on the number of people you are addressing. Your talk will not look the same if you are presenting to two people compared with 200. A presentation to a large audience must be more 'theatrical' in the delivery, amplifying body gestures and voice, for example. Speaking to a large public will also push you towards using humour more.

With smaller audiences, the same talk could seem out of place or even uncomfortable. Standing up to speak to two people is not always a good idea. In front of a more substantial audience it is obligatory, so that they can all see and hear you clearly. Handing around a sample to a meeting room with only five people is possible, but in an amphitheatre, it is more likely to create havoc. You need to make sure well in advance that your exercises, demos, or handouts are suitable for the group size you are presenting to.

## 3.2.2  Interaction

When giving a presentation, yes, you are the centre of attention. You will be talking to your audience for the majority of the time. However, there will no doubt be a moment where people in your audience can get involved too. The level and importance of that interaction will depend on the context of your talk. You should know in advance what kind of audience contact to expect. The best improvisers come prepared.

How you give your talk will be fundamentally different based on the way you interact with your spectators. The most common scenario is question-response contact. In a boardroom of five people, they can simply cut you off to ask a question, which is much less likely in a packed auditorium of several hundred spectators where queries wait until the end.

At the end of your talk at a conference for example, a chairman will often be present who will open the floor to questions. This means that anybody sitting in your audience can stand up and interrogate you. Whilst you may sometimes be given these questions beforehand, it is much more common that you will be required to improvise. We recommend you prepare responses to any important questions you think may come up.

The technical level of these questions varies with the proportion of specialists in the room. For example, you will have to answer more expert questions after giving a talk at a conference in your field, than when talking to teenagers in a high school. In other cases, your audience may be assessing or judging you, such as in a thesis defence. In those instances, the questions will be purposefully difficult in order to challenge you and can lay at the very border of your knowledge base. In some circumstances, there is no allotted time for audience questions. Any exchanges with your audience will happen 'behind the scenes', with a smaller group of people after your presentation.

When taking visitors around a site on a tour, there will possibly be a very loose flow of questions and responses, back and forth. The same goes for a seminar or workshop. These questions are more informative and will not be meant to test you. They may even influence how you deliver your pre-prepared talk, as you adapt to the needs and curiosities of your audience. There is no harm in improvising, but at least go into your game with a robust plan that you can adapt if necessary.

## 3.2.3  Duration

You should know how much time you have and stick to it—in some cases, to the second! Running over your time slot, at a conference for example, is frustrating for your audience, who may want to dash off to another talk after yours. It is also disrespectful to other speakers coming after you who then have less time for their own presentations. Conversely, whilst potentially a little embarrassing, cutting your talk short because you've run out of content, say during a plenary session, is far better than rambling on for an extra half an hour. Especially since there are almost always audience questions to fill the time.

We advise that you should aim to run your talk time to just under your allotted slot duration. This rule is particularly relevant for situations where you have a strict time limit, such as in a pitch for funding or a competition. In cases like these, you may even be penalised for running over. Certain formats where time management rules may differ include speaking on camera, as the footage can be cut down. But even then, if you talk for too long, your speech will be cut up and diluted—*but not by you*.

Talks can be of any length, generally depending on restrictions of the event where you will be speaking. For simplicity, we have divided potential presentation durations into short, medium, long and variable. There could be some cross-over between groups.

### 3.2.3.1  Short

Don't be fooled. Preparing a short presentation does not take less time than a long one. Usually it is the opposite, because timed talks require very rigorous planning. The shortest science presentation that you will probably ever do is the three-minute format, but they can be shorter. We consider short as anything that is under 10 min.

Commonly, presentations like these give the opportunity for many people to talk at the same event. There are two main reasons for this. First, it allows a time limit for assessment and comparison of talks during competitions, such as Three Minute Thesis and pitches for funding. Second, it gives the chance for numerous people to promote or expose their work in a fair way. This includes a series of short talks at a ceremony, for example. Either way, strict time management is crucial so you will need to focus your talk on a very specific aspect of your topic.

Short talks are often, but not always, intended to present science to a nonspecialist audience. Clarity, efficiency and standing out from the crowd are paramount to succeeding in presentations of this type.

### 3.2.3.2   Medium

The average length for a science talk is somewhere between 15 and 30 min. If you are invited to speak as an expert to your peers, as a guest speaker at a lab or meeting, this is the length of time you will probably be asked to talk. You should be describing your work in enough detail to argue for your contribution to your field, whilst still having a clear, memorable message. This type of presentation is generally followed by a reasonable question/answer session or debate.

Opportunities to share your work with other specialists or a technical audience is a great way to influence your field and get feedback about your research. For this reason, presentations of this length can be extremely nourishing for your own projects and help you with ideas on how to improve or move forward.

### 3.2.3.3   Long

In this book, we consider a long talk to be anything over 30 min. The most common ones will be around an hour and the content level can vary. This category includes a plenary session at a conference or a talk for the general public. Such a long presentation will generally be by invitation only and you should be very honoured. Time is a precious commodity and if an event organiser believes you deserve so much of it, then take it as a great sign of respect for your expertise.

As such, you should also respect the time that you have been given. Your audience will expect you to be at the very top of your game. If you are talking to experts in astrophysics, adapt your presentation to them. If they are the general public, keep them on their toes and full of curiosity until the end.

### 3.2.3.4    Variable

Whilst it is uncommon that you will not have a fixed length on a presentation, it does sometimes occur. This includes poster sessions at conferences or guests in your lab. It may even be a visit by a film crew who are interviewing you for a documentary. In this context, they will probably leave their time slot relatively wide to help you adapt your talk and get used to the camera. Don't be fooled, however: you should normally have received questions in advance of the shoot date and have prepared your answers the same way you would for a short talk.

## 3.2.4    Your Stage

The space provided for you to give your talk is a component to consider when preparing. It will influence what you can do and how you will stand or move whilst presenting. You may sometimes have nothing but a lectern to stand behind, or you may have the whole floor at the bottom of an amphitheatre to play with. A typical TED talk, for example, will normally take place in a theatre setting, with movement limited to the edges of the famous round, red rug that is placed somewhere in the centre of the stage. Whether it is a spot at the front of a meeting room or the podium at Wembley Stadium, in this book we refer to this as your stage.

### 3.2.4.1    Layout

Not all spaces where you will give a talk or presentation look the same. When we support speakers before a high-impact event, we provide them with the maximum amount of information about their stage in advance. When possible, we show them pictures of the space or, better still, we visit before the actual day. The reason for this is that the layout of the room can have a huge impact on the delivery of your presentation.

Firstly, it can affect the atmosphere and thus your audience's mood. Certain spaces are solemn because they have a particular history, like a gloomy library, whereas others may be furnished with sofas and comfy seats, providing a very pleasant, approachable feeling. Bright or gloomy, narrow or wide, the atmosphere of a space can change the mood of the room, so try to be aware of this beforehand.

Another point to consider is the way in which the audience is facing you. If your audience is sitting comfortably, like in an auditorium, they can listen to you for many minutes without distractions. During a meeting, your colleagues may be sat in a U-shape around a table. With this setup, you should probably stand and move around to ensure that your whole audience sees you. For a visit of a research centre, a poster session or a science festival, your audience could find themselves stood up around you. It is in your best interest to position them in a place where they can all see and hear you well. They may need time to rest or move around too. If you do ever give your presentation outdoors, you should take weather and background noise into account during your preparation as well.

### 3.2.4.2  Equipment

Do you need a projector for your PowerPoint slides? A whiteboard to draw on? A flipchart for brainstorming? Most meeting rooms, for example, give you access to these things, or you can request them. Don't leave it until the day of your presentation to reserve a clip-on microphone or find a room equipped for videoconferencing, only to discover that they are all booked out for another event. To make sure that the correct equipment is accessible, find out as early as possible what should be there, how to reserve extra materials you might need and who to call if it isn't available when you arrive.

In auditoriums, you can be lucky as there might be a technician to hand who can help you with technical faffs, like testing microphones, screens, projectors or even lighting. They are generally in-house and will know their way around their own kit. However, do your best to test everything beforehand and give the technician explicit indications of what your expectations are.

## 3.3  Occasion

There is a long list of presentation types that you may come across, but in reality, they all tend to fit into categories based on the occasion during which you will be doing your talk. Each one has rules—implicit or explicit—that you need to know in advance. So, in this section, we will take you through a whistle-stop tour of the main occasions you will come across.

### 3.3.1  Meeting

During a meeting, you will generally be presenting to people who already know you and your research topic. Your audience is there to find out more about the advancement of your research or to make a decision based on your data. You will likely be allocated between 10 and 30 min for your presentation, but it can run over. Due to the close proximity with your audience in this instance, you will commonly be interrupted with questions throughout your talk, rather than at the end.

### 3.3.2   Conference

Speaking at a conference is a sign of recognition. If you have been invited to talk in front of a large audience, you are an acknowledged expert in your domain. At a research conference or a more general event, you will commonly be given between 15 min and one hour to talk, followed by a question/answer session. This type of presentation allows you to cover several aspects of your research in detail and for a targeted audience. Often, you will be introduced by a chairperson who will read your biography to the audience just before you speak. And, if you go over the allocated time slot, their role is to interrupt you. You should be vigilant in respecting this restriction. See more about this in Chap. 12.

### 3.3.3   Site Tour

As a researcher, you will often invite people into your laboratory: partners, students, collaborators, clients or even politicians. Visits provide a special moment to present your facilities and to showcase your research. A well-prepared visit follows a pre-defined itinerary; usually starting with a presentation of the research institute, followed by short-talks or demonstrations given by the different laboratories that may be of interest to your visitors. With guests being stood up for a large part of the visit, we recommend limiting the duration of short talks. Anywhere over 20 min and your visitors will switch off—keep them engaged to benefit as much as possible from this experience.

### 3.3.4   Poster Session

A poster session is a great opportunity to show off your research to the world, mainly at conferences. The goal here is to reel in fellow experts to get a conversation going about your work. You can prepare for a poster session in the same way you prepare for any other talk. You should have a pitch ready for any passers-by so that you can take them through your poster quickly and effectively. You should also prepare yourself to answer precise questions about your work—especially the methods and results sections. The whole point of doing a poster is so that you can discuss your topic with people face-to-face and one-on-one. They can be extremely enriching moments because they are a great opportunity to develop your professional network, increase the visibility of your projects and share ideas with other experts.

### 3.3.5 Thesis Defence

After three, or even four, years of hard work and months of rigorous dissertation writing, you are likely to be worn out by the time you start preparing your thesis defence. Try to save some time and energy for your presentation. Depending on the country where you are based, or your research topic, thesis defences may occur in public or behind closed doors. Either way, you will have a jury whose job it is to critique your work and ask you questions. Even in the case where your defence is public, your target audience should always be the jury. They are experts in your field and your job is to convince them that you are too. As such, do not oversimplify your talk for your Gran sat in the audience (she will be impressed with you either way!). Be prepared to answer tough questions that require thought. If it helps, you can keep in mind that any talk you give afterwards will never be criticised as heavily, so will always be more enjoyable.

### 3.3.6 Prize Ceremony

Congratulations! You won a prize! Whilst actors spend their acceptance speeches at the Oscars thanking their mums, dads, agents and co-workers, scientists usually don't. Winning an award, grant or prize is fantastic not only for gratification, but also for the visibility it can give you and your research. At a science award ceremony, you will generally be given a moment to present your project. The audience will likely be filled with fellow scientists, stakeholders, decision-makers and other important people who can influence the future of your research. Take this opportunity by the horns, show them why your work is amazing.

### 3.3.7 Competition

During a competition, such as FameLab or Three Minute Thesis, speakers are often invited to make a short three-minute pitch-type presentation. If you ever participate in one, you should be aware that you will be evaluated by a jury and/or the public. There will be a winner who is chosen based on a number of different pre-determined criteria: impact of the research topic or innovation presented, quality of oral presentation and even likeability.

### 3.3.8    Camera

Some of the above public speaking opportunities may be filmed. This information is useful to know in advance to avoid being caught with your pants down on the day of your talk. The presence of cameras can be a source of stress for some people. Being mentally prepared for this eventuality will help you avoid falling to pieces on the day. On the upside, no specific additional preparation is required for this type of occasion. Simply, be aware that putting a little extra effort into combing your hair and ironing your shirt will be worth it—the videos will be circulating for a while after the event, so you will feel better about yourself if you look presentable.

### 3.3.9    Interview

We will not discuss in detail how to deal with interviews because it is not the subject of this book. What we will say though is that even if an interview is a different kettle of fish than public speaking, we highly recommend that you prepare ahead of meeting a journalist. Often you can request to see the questions they plan to ask you in advance. Otherwise, we advise you to prepare a list of those which are likely to be asked and your answers. Be ready for trick questions too. The journalist will only use a few minutes, or even seconds of your interview. Make sure that what you say appropriately reflects your thoughts and be concise.

## 3.4    Logistics

There is a final type of preparation that is worth getting right, which involves reducing anxiety to a minimum. The more comfortable you are when the day comes, the better your talk will be. Asking as many of the right questions as possible in advance of the day will mean that no last-minute panics will ruin your zen, and consequently your talk.

We suggest using the what, who, where, when, how, why and how many method (also referred to as the WWWWHWH method) to ask yourself the right questions. Here are a few to get you going.

### 3.4.1  What?

Questions involving objects and any materials:

- What kind of room will it be/should I pick?
- What equipment will be available in the room (projector/screen/board/microphones)?
- What do I need to provide the participants (pens/paper/sticky notes)?

### 3.4.2  Who?

Questions to do with people:

- Who is organising the presentation?
- Who is inviting the audience?
- Who are the other speakers?
- Who is responsible for technical logistics?
- Who is in charge of catering (drinks/meals)?
- Who pays for the expenses?

### 3.4.3  Where?

Questions related to the location of your presentation:

- Where is the presentation taking place? Where is the building? Where is the room?
- Where will people eat?
- Where will the participants be staying?
- Where can I park?

### 3.4.4  When?

Questions involving duration, schedule or frequency:

- When is my talk?
- When should I arrive?
- When will my audience arrive?
- When is the break?
- When should I start/finish?

## 3.4.5 How?

Questions related to means and methods:

- How is the furniture in the room arranged (e.g. U-shaped, round table or face-to-face)? What is the layout of the room (e.g. amphitheatre or conference room)?
- How will the participants enter the stage?
- How will I be introduced?
- How can I hand out documents during the meeting?
- How do I switch on the video projector?
- How do I get into the room? Do I need a key? Will there be someone to meet me when I arrive?

## 3.4.6 Why?

Questions related to reasons to give a presentation:

- Why am I giving this speech?
- Why is it an oral presentation?
- Why are people attending this event?

## 3.4.7 How Many? How Much?

Questions to do with quantities:

- How many participants will attend?
- How much will it cost?
- How many handouts do I need to bring?

# 3.5 Moving on…

Without a doubt there are tonnes more questions that you can ask yourself, but this list should get you going. Feel free to add or remove as necessary. Once you feel that you have scoped out the context of your presentation, you're ready to move onto the next chapter to define the message of your talk.

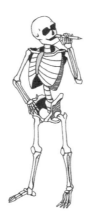

# 4

## Define Your Message
### Get to the Point of Your Talk

When did you last hang a picture frame on a wall? Did you use a hammer and a nail? Picture that in your mind for a second. You took the nail out of the toolbox, placed it to the wall and with the other hand hit it with a hammer. With the first smash, the nail no doubt entered the wall a little, yet you probably still had to keep hold of it to maintain its position. Then, as you hit over and over again, the nail will have gradually become entrenched deeper and deeper until it was sturdily jammed in the wall.

You need to see your message as that nail. Your objective is to use your presentation as a hammer, to keep banging until it becomes irrevocably wedged in the minds of your audience. Effective public speaking gets a point across to the audience as efficiently as possible.

Previous British Prime Minister, Winston Churchill, was renowned for his speeches and is often described as one of the best public speakers of the

© Springer Nature Switzerland AG 2020
A. Youknovsky and J. Bowers, *SELL YOUR RESEARCH*,
https://doi.org/10.1007/978-3-030-34181-7_4

20th Century. When talking about how to produce a good speech, he once famously said:

> *If you have an important point to make, don't try to be subtle or clever. Use a pile driver. Hit the point once. Then come back and hit it again. Then hit it a third time - a tremendous whack.*

What he refers to as a "point", in communication we refer to as a message. In public speaking, you should know exactly what you want your audience to remember and hand it to them on a platter. Science presentations follow the same rules. Your job as a presenter, speaker or facilitator is to make sure that happens.

By now, you should be aware of the context surrounding your talk. You should know what you are doing, who your audience is and, at least a little bit, about their motivation for attending your talk. In this chapter, we take you through the steps of how to effectively combine your own objective and the expectations of your audience to define your message. We provide several different examples in the form of case studies, which are adapted from our own personal experience.

## 4.1   Four Questions to Define Your Message

In Chap. 2 we described the communication model. When communicating, you are transferring a precise piece of information to your audience: your message. To be able do this, you should start by correctly defining what that message is. Once you have a clear idea, you can then structure your presentation in such a way that you successfully facilitate its transmission.

Your message should be well adapted to your audience. To define it, you need to take their expectations into account. This means that even with the same personal objective, your message will change depending on who you are speaking to.

When communicating, people often get lost in the trap of talking about 'me, myself and I'. But your audience doesn't care that much because all they are thinking about is 'me, myself and I', too. In France, a famous song by Jeanne Moreau begins, "Parlez-moi de moi, y'a que ça qui m'intéresse." Roughly translated, it means, "Tell me about myself, that's all I care about". This is how you need to see your audience.

They came to find out how your work is relevant to their lives. You need to put your audience at the centre of your thought process. Hence, your message needs to explicitly describe the importance of your presentation to them. To define it, whilst considering your audience, we advise that you answer the following questions in this order:

1. What is your objective?
2. What are your audiences' expectations?
3. What arguments do you have to support your objective with this specific audience?
4. What is your message (in one sentence)?

Your responses to the first two questions, once combined, will allow you to answer the third, where you can begin to accumulate your arguments. By selecting the strongest of these arguments, you can start to build your message. It is sometimes very wordy at this point and may be two to three sentences long. Therefore, we propose a final step of transforming it into a shorter, sexier, single sentence: your message. You will use this as the foundation for your talk.

## 4.1.1  What Is Your Objective?

Firstly, you need to define your objective starting from your own perception. Your objective is what *you* want to get out of the talk. You know the context of your talk, but you need to think here about what is motivating you to speak about that part of your research. Your objective should be a measurable act: what do you want your audience to think? What do you want them to do?

Examples include:

- I want my audience to think that it is possible to live in a world without plastic.
- I want my audience to fund my electron microscopy start-up company.
- I want my audience to collaborate with me on a European water treatment project.

## 4.1.2  What Are Your Audiences' Expectations?

Next, you need to think about the motivations of your receivers: what drove them to attend your talk. Take that into account. And if you are not familiar with your audience, you should inquire about them in advance. Your goal is simple: talk to your spectators about what interests them.

A manager may want to know why he should hire that technician you requested. An investor may want to know what their returns on investment will be. A thesis jury member may want to know what is unique about your research. Each type of person will be expecting you to deliver something to them. In doubt, go back to Chap. 3 to find a list of audiences and their possible motives.

It is not always a simple task to try and get into the heads of your planned audience. Completing the following sentence could help:

> As (title/skill of people in your audience), what I am expecting from this presentation is…

Your response could be as simple as, "…to know my risk of heart disease and what I can do to protect myself." Whoever your audience is, you should be able to break down their expectations into a few sentences.

### 4.1.3   What Arguments Do You Have to Achieve Your Objective with This Specific Audience?

Now, this is where we start getting into the nitty gritty. You need to find arguments you can offer your audience to achieve your objective. Imagine you are giving your talk at a meeting with a funding body with the following objective: "I want my audience to fund my research project on metabolism in microalgae." In this case, your arguments could include:

- Microalgae are important for the environment as they produce 50% of the oxygen in the air we breathe.
- By understanding metabolism in microalgae, we could find solutions to deal with the consequences of climate change.
- My team of researchers are proficient in modern analytical techniques required to carry out this project.
- Last year, we published six peer-reviewed articles on the topic.
- My research topic is multi-disciplinary: biophysics, genetics and marine biology.

### 4.1.4   What Is Your Message?

Your message, or "point" as Churchill referred to it, is a combination of your strongest arguments, which are best adapted to your audience. There may be several that fit the bill. However, you need a message that is concise, memorable and easy to understand. What do you want people leaving your talk to have at the forefront of their minds? For this section, you should try to finish this sentence:

*The one thing that I want my audience to remember from my talk is…*

Pick arguments that you feel are the most directly relevant to your audience. You should then work to make the message slicker. The words you use here need to be precise and fit into one simple sentence. Where possible, use facts, figures and imagery that the audience can relate to.

Here are some examples:

- This data mining software will allow emergency services to detect online terrorist activity within a matter of minutes.
- You will improve the success rate of your gastric bypass surgeries by 20% using our innovation.

### 4.1.5    Putting It into Action

There goes the theory. Applying it needs some practice. Don't be afraid, you will get used to it. For the rest of this chapter, we provide you with concrete examples of how to define a message. You should work through these illustrations and bear in mind how to apply the message-defining method to your own research presentations.

## 4.2    Case Study I: The Headteacher

We will start with a simple, non-scientific example to give you an idea of how our method works. A headteacher at a specialised school for children with hearing impairments is holding an open evening. She would like to introduce the school to new parents and convince them that they should send their children to her establishment in the upcoming school year. She will have a dedicated moment when all of the parents are gathered to make a speech.

Believing strongly that her teaching staff are of a high calibre, she would like to use that as a selling point in her talk. She also has facts and figures representing the exam grades of previous years. In this case study, the speaker has a clear objective for her talk. She wants parents to send their children to her school. She knows the context of her speech and should have a good idea of the people in her audience.

She wants to use her teaching staff as an argument to defend her main objective. This is understandable. But what about her audience? What are they interested in? Well, parents are more pre-occupied with their children. The teachers' abilities are a means to help their children give their best, but they are not an end in themselves. If the headteacher talks about her team, there is a chance that the parents will not be convinced that their child is fully considered. The parents attending the open evening are expecting the best education for their child. This is what the headteacher should talk to them about.

We propose a message which is adapted to the parents' expectations by putting their children at the centre of it. This message uses certitudes like "your children will..." and emphasises positive results, like "... achieve their full potential." In Table 4.1, we summarise these responses to the questions.

This case study is a very general example of how you can adapt your message to an audience. Here, we have made the choice to focus the message on the teaching staff. Why is that so? If your message resembles a long list of arguments, it will hardly be remembered. Try to focus on your strongest arguments. What makes your project so special for that particular audience? Tell them about that.

However, this doesn't mean that she cannot include her other arguments in her presentation. On the contrary, high exam grades are proof that the children in her school are reaching their full potential and good facilities will help both teachers and children to work well. These points serve to defend her overall argument even if they are not specifically included in her message.

## 4.2.1 Science Talks

If you are reading this book, then your messages will almost always be based on your scientific or technological research. Here, we look a little deeper at how to define your message in the context of a scientific or technical presentation. Looking back at Chap. 3 we discussed the different types of audiences of science presentations. You should be aware of who is in your audience, and why they are there. Now, let's look at another case study, taking into account two audience types you may come cross in your talks: experts and non-experts.

**Table 4.1** The headteacher's message

| | |
|---|---|
| WHAT IS YOUR OBJECTIVE? | MY OBJECTIVE IS... <br><br> ...FOR YOU TO REGISTER YOUR CHILDREN AT MY SPECIALISED SCHOOL FOR HEARING-IMPAIRED CHILDREN. |
| WHAT ARE YOUR AUDIENCES' EXPECTATIONS? | AS A PARENT, WHAT I AM EXPECTING FROM THIS PRESENTATION IS... <br><br> ...TO FIND OUT WHETHER MY CHILD WILL BENEFIT FROM THE BEST EDUCATION AT THIS SCHOOL. |
| WHAT ARGUMENTS DO YOU HAVE TO ACHIEVE YOUR OBJECTIVE FOR THIS SPECIFIC AUDIENCE? | MY MAIN ARGUMENTS ARE... <br><br> ...WE HAVE A FANTASTIC TEAM OF TEACHING STAFF, <br> ...WE HAVE A PROVEN TRACK RECORD OF HIGH EXAM GRADES, <br> ...WE HAVE FACILITIES ADAPTED TO THE NEEDS OF HEARING-IMPAIRED CHILDREN. |
| WHAT IS YOUR MESSAGE? | THE ONE THING THAT I WANT MY AUDIENCE TO REMEMBER FROM MY TALK IS... <br><br> ...THAT YOUR CHILDREN WILL ACHIEVE THEIR FULL POTENTIAL AT OUR SCHOOL THANKS TO THE QUALITY OF OUR TEACHING STAFF. |

## 4.3    Case Study II: The Bionics Professor

This example is adapted from an excellent TED Talk by Hugh Herr, a biomechatronics professor at MIT. Prof. Herr, himself a double amputee, leads a team who design bionic limbs, which are specifically adapted to the human physiology. They later published a research paper presenting a bionic ankle joint that they had modelled on a biological one. The particularity of this innovation was that they specifically recreated the ankle joint of a dancer. They tested the prosthetic on Adrianne Haslet-Davis, who had been a professional dancer until she lost the bottom half of her left leg in the Boston Marathon bombing.

In this case study, we use a hypothetical example where Prof. Herr gives two talks to present his findings. One at a conference with his peers and the other as a TED Talk. As we mentioned, his message will differ depending on the audience.

### 4.3.1    Experts

Let's imagine Prof. Hugh Herr is invited to a conference organised by the International Bionic Limb Society. He wants to reveal his innovative design for a bionic ankle joint, which he claims is closer to imitating biological movement than any other previous limb design, particularly in terms of dance. People who attend this conference are primarily scientists who carry out research in bionic limb design: medical professionals, mechanical engineers, biomechanics experts and so forth. Experts are looking to learn something useful for their own work, or to foster new collaborations. Table 4.2 describes how he could have defined his message using our four-question system.

### 4.3.2    General Public

Now, we know this professor of bionics was invited to do a TED Talk about the very same research. The audience is non-specialists; people who are attending the talk because they are curious about new, innovative ideas. A TED Talk audience tends to be educated adults of various ages and professions. They want to be inspired and find out how the topic will impact them or society. The outline of how we suggest he build his message can be found in Table 4.3.

**Table 4.2** The bionics professor's message to experts

| | |
|---|---|
| WHAT IS YOUR OBJECTIVE? | MY OBJECTIVE IS... <br><br> ...TO FIND OTHER RESEARCHERS WHO WANT TO ADAPT MY ANKLE JOINT DESIGN TO OTHER LIMBS. |
| WHAT ARE YOUR AUDIENCES' EXPECTATIONS? | AS A FELLOW EXPERT IN BIONICS, WHAT I AM EXPECTING FROM THIS PRESENTATION IS... <br><br> ...TO DISCOVER IN WHAT WAY THIS SCIENTIST'S RESEARCH PROVIDES NEW INSIGHTS IN THE FIELD OF BIONIC LIMBS/ANKLES, <br> ...TO ADVANCE MY OWN RESEARCH WITH NEW IDEAS. |
| WHAT ARGUMENTS DO YOU HAVE TO ACHIEVE YOUR OBJECTIVE FOR THIS SPECIFIC AUDIENCE? | MY MAIN ARGUMENTS ARE... <br><br> ...THIS NEW BIONIC LIMB CAN ACHIEVE THE SAME TASKS AS BIOLOGICAL ONES FOR DANCING, <br> ...WE ACHIEVED SIGNIFICANT IMPROVEMENT IN MOVEMENT COMPARED TO TRADITIONAL PROSTHETICS, <br> ...THIS HAS NEVER BEEN DONE BEFORE, <br> ...THE PROCESS ONLY TOOK 100 DAYS, <br> ...THERE ARE OTHER APPLICATIONS/ LIMBS TO WORK ON. |
| WHAT IS YOUR MESSAGE? | THE ONE THING THAT I WANT MY AUDIENCE TO REMEMBER FROM MY TALK IS... <br><br> ...BY USING OUR TECHNIQUE, YOU TOO CAN DESIGN NOVEL PROSTHETIC LIMBS IN UNDER 100 DAYS. |

**Table 4.3** The bionics professor's message to non-experts

| | |
|---|---|
| WHAT iS YOUR OBJECTiVE? | MY OBJECTiVE iS... THAT YOU UNDERSTAND THE iMPORTANCE OF DESiGNiNG BiONiC LiMBS CAPABLE OF EFFECTiVELY iMiTATiNG THE MOVEMENT OF BiOLOGiCAL ONES. |
| WHAT ARE YOUR AUDiENCES' EXPECTATiONS? | AS A NON-EXPERT, TED-TALK SPECTATOR, WHAT I AM EXPECTiNG FROM THiS PRESENTATiON iS... ...TO FEEL iNSPiRED BY A GREAT SPEAKER ...TO LEARN THiNGS ABOUT A SCiENTiFiC DEVELOPMENT THAT I PREViOUSLY KNEW NOTHiNG ABOUT. |
| WHAT ARGUMENTS DO YOU HAVE TO ACHiEVE YOUR OBJECTiVE FOR THiS SPECiFiC AUDiENCE? | MY MAiN ARGUMENTS ARE... ...BiONiC LiMBS, CAPABLE OF EFFECTiVELY iMiTATiNG MOVEMENT OF BiOLOGiCAL ONES, OFFER NEW POSSiBiLiTiES FOR PEOPLE WiTH DiSABiLiTiES, ...THERE iS A LOT OF SCiENCE BEHiND THESE iNNOVATiONS, ...WE HAVE DESiGNED AN ANKLE JOiNT AND TESTED iT ON DANCERS—AND iT WORKS! |
| WHAT iS YOUR MESSAGE? | THE ONE THiNG THAT I WANT MY AUDiENCE TO REMEMBER FROM MY TALK iS... ...NEW BiONiC TECHNOLOGiES CAN RESTORE THE PHYSiCAL ABiLiTiES OF AMPUTEES. |

### 4.3.3  Different Messages

For specialists, the point to drive home is how the research is different and in what way it is new information. For the general public, the goal is more about how this innovation will change their lives or impact the wider world in some way. You will also notice that the message for the public is broader. Experts will know more about the topic and have the background understanding of the impact of your very precise finding. On the other hand, the non-specialist audience needs to know the bigger picture.

In the real version of this example, Hugh Herr goes further with this in his TED Talk. He cashed in on the emotional impact of his discovery by calling the ex-dancer amputee onto the stage. Relying on the new bionic limb that Hugh's team created, she danced again for the first time since her accident, live on stage during his talk. Whilst this emotional style of talk is not suited to every audience, this demonstration, coupled with Adrianne's emotion after her dance, made it particularly emotive and memorable.

## 4.4  Case Study III: The Entrepreneur

Our final example looks at a hypothetical engineer turned entrepreneur. She has set up an eco-friendly spin-off company from her lab, which aims to recycle waste generated by coffee machines. Her system converts used coffee grounds into logs, which can then be burnt as fuel for stoves. Along with reducing waste, her goal is to provide an alternative to cutting down trees or burning coal for wood stoves.

In her business plan, our entrepreneur aims to collect used coffee grounds from cafés free-of-charge. Hence, both recycling and obtaining her starting material for no cost. The company will then convert it into logs, to be sold for a profit. Let's imagine that this engineer is trying to launch her company. To do so, she has managed to put three presentations in place, each with different groups of individuals in the audience: investors, coffee shop managers and end-users (clients). How will she adapt her message?

### 4.4.1  Investors

Anyone looking to invest or fund a project will expect a return of some sort. These people need to know why you want the money, how you plan to spend it and how they will benefit from it. In our example, the entrepreneur wants to convince private investors to give her enough money to run the company for a year in exchange for an 8% return on investment from the second year (see Table 4.4).

### 4.4.2  Coffee Shop Managers

The entrepreneur also needs coffee shop managers to provide her with used coffee grounds free-of-charge. Whilst it is great for the environment, it is inconvenient for managers to have to separate their litter into coffee waste and non-coffee waste. Practicality often trumps sustainability in business models. To get around this barrier, the entrepreneur has designed a reward scheme for coffee shop managers who help her. She creates a label, which can be displayed on the front of coffee machines or bistros indicating, "Here, we recycle our used coffee," whilst specifying that it is good for the planet. Something like, "Committed to recycling coffee waste." (see Table 4.5).

The entrepreneur will give details in her presentation about the recycling process. However, the simplicity of the procedure does not seem to us to be the highest priority argument to highlight in her message. On the other hand, if you believe that your audience may be unhappy about adopting a new technology, and that this would be a major blockage in your project, this argument could be brought to the forefront.

### 4.4.3  Clients

The entrepreneur makes a presentation to an audience of potential clients at an exhibition for sustainable products. Like any client, they want a good value, high quality product. She hopes to convince them that they should buy her logs. Her clients are eco-conscious and interested in buying sustainable goods. But they must also be reassured of the quality of her logs, which need to be at least as good as ordinary firewood. They also need to know that these coffee-based logs will not damage their wood stoves or require any costly additional material (see Table 4.6).

## 4.5  Messages for Mixed Audiences

We cannot ignore the possibility that you may be in a situation where you have an audience comprised of several different categories of people. In these cases, your audience will be mixed and therefore it is more difficult to identify the message. When is this likely to happen? At a conference, you will have varying levels of experts in your field. For example, if you manage to

**Table 4.4** The innovative entrepreneur's message to investors

| WHAT iS YOUR OBJECTiVE? | MY OBJECTiVE iS... ...THAT YOU iNVEST iN MY SPiN-OFF COMPANY. |
|---|---|
| WHAT ARE YOUR AUDiENCES' EXPECTATiONS ? | AS AN iNVESTOR, WHAT I AM EXPECTiNG FROM THiS PRESENTATiON iS... <br>...TO KNOW HOW MY MONEY WiLL BE SPENT, <br>...TO UNDERSTAND WHAT THE RETURN ON MY iNVESTMENT iS LiKELY TO BE, <br>... TO FEEL REASSURED THAT THE COMPANY'S TEAM WiLL ACHiEVE THE OBJECTiVES OF THE PROJECT. |
| WHAT ARGUMENTS DO YOU HAVE TO ACHiEVE YOUR OBJECTiVE FOR THiS SPECiFiC AUDiENCE? | MY MAiN ARGUMENTS ARE... <br>...OUR COMPANY iS PROViDiNG AN ENViRONMENTALLY FRiENDLY TECHNOLOGY, <br>...THE LOGS WiLL COST THE SAME AS 'NORMAL' LOGS, FOR THE SAME QUALiTY, <br>... WE WiLL CREATE 20 JOBS iN ONE YEAR, <br>...WE ARE OFFERiNG YOU 8% RETURN ON iNVESTMENT FROM THE SECOND YEAR. |
| WHAT iS YOUR MESSAGE? | THE ONE THiNG THAT I WANT MY AUDiENCE TO REMEMBER FROM MY TALK iS... <br>...BY FUNDiNG OUR ECO-FRiENDLY COMPANY, YOU WiLL EARN AN 8% RETURN ON iNVESTMENT FROM THE SECOND YEAR. |

**Table 4.5**  The innovative entrepreneur's message to coffeemakers

| WHAT IS YOUR OBJECTIVE? | MY OBJECTIVE IS...<br><br>...TO CONVINCE YOU TO GIVE ME YOUR USED COFFEE GROUNDS FREE OF CHARGE |
|---|---|
| WHAT ARE YOUR AUDIENCES' EXPECTATIONS? | AS A COFFEEMAKER, WHAT I AM EXPECTING FROM THIS PRESENTATION IS...<br><br>...TO FIND OUT WHAT BENEFIT THERE IS FOR ME IN RECYCLING AND GIVING YOU MY COFFEE WASTE,<br><br>...TO UNDERSTAND THE WHOLE PROCESS AND VERIFY THAT IT IS SIMPLE. |
| WHAT ARGUMENTS DO YOU HAVE TO ACHIEVE YOUR OBJECTIVE FOR THIS SPECIFIC AUDIENCE? | MY MAIN ARGUMENTS ARE...<br><br>...IT IS GOOD FOR THE ENVIRONMENT,<br><br>...YOU WILL RECEIVE AN OFFICIAL 'ECO-FRIENDLY' LABEL TO DISPLAY TO YOUR CUSTOMERS,<br><br>...MANY CUSTOMERS ARE REASSURED BY BRANDS WHO DEMONSTRATE COMMITMENT TO ENVIRONMENTALLY FRIENDLY SOLUTIONS,<br><br>...THE RECYCLING PROCESS IS SIMPLE. |
| WHAT IS YOUR MESSAGE? | THE ONE THING THAT I WANT MY AUDIENCE TO REMEMBER FROM MY TALK IS...<br><br>...BY RECYCLING YOUR USED COFFEE, YOU WILL RETAIN YOUR EXISTING CLIENTS AND GAIN NEW ONES. |

**Table 4.6**  The innovative entrepreneur's message to end-users

| WHAT IS YOUR OBJECTIVE? | MY OBJECTIVE IS… <br><br> …TO CONVINCE YOU TO BUY MY RECYCLED COFFEE LOGS. |
|---|---|
| WHAT ARE YOUR AUDIENCES' EXPECTATIONS? | AS A CLIENT, WHAT I AM EXPECTING FROM THIS PRESENTATION IS… <br><br> …TO DISCOVER A NEW ENVIRONMENTALLY FRIENDLY LOG, <br><br> …TO CHECK THAT THEY ARE AS EFFECTIVE AS THE PRODUCT I AM ALREADY USING FOR THE SAME PRICE, <br><br> …TO BE REASSURED ABOUT THEIR QUALITY BECAUSE I DON'T WANT TO DAMAGE MY WOOD STOVE. |
| WHAT ARGUMENTS DO YOU HAVE TO ACHIEVE YOUR OBJECTIVE FOR THIS SPECIFIC AUDIENCE? | MY MAIN ARGUMENTS ARE… <br><br> …OUR LOGS ARE NOT MORE EXPENSIVE, <br><br> …THEY WORK JUST AS WELL AS MARKET COMPETITORS, <br><br> …THEY ARE ECO-FRIENDLY BECAUSE THEY RE-USE WASTE AND ARE LOCALLY SOURCED, <br><br> …OUR PROJECT WILL CREATE 20 JOBS IN YOUR REGION. |
| WHAT IS YOUR MESSAGE? | THE ONE THING THAT I WANT MY AUDIENCE TO REMEMBER FROM MY TALK IS… <br><br> …YOU CAN STAY WARM WHILST PRESERVING THE PLANET. |

bag yourself a plenary session then your talk will be attended by people from all around the conference: other scientists, communicators, journalists, salespeople etc. The same goes for exhibitions.

Your audience will also be mixed if you ever have an open day at your lab. You may have people coming from both inside and outside your institute, meaning that your audience will range from expert to non-scientist level. Your audience members will also vary in profession, age, gender, education-level, personal interests and so on. Depending on the situation, there are two ways to adapt your message to a mixed audience: prioritise your audience or find a common link between them.

### 4.5.1　Prioritise Your Audience

Whilst your audience may be mixed, you can choose which members to target the most. Imagine you have funders, colleagues and general public in an audience for your short talk. Your goal may be to secure more funding for your research for the next three years. In that case, your priority should be the funders. Your message should target them. But maybe go easy on the details about your business plan—if your other spectators start snoring, it will become distracting. On the other hand, if your objective is to inspire more people to train to be scientists, then you should prioritise those audience members who are in the general public category.

### 4.5.2　Find the Common Link

If you cannot prioritise the groups in your audience, then you need to find a connection between them. Let's take our case study of the entrepreneur. What happens if her three audiences are mixed? Perhaps the entrepreneur is talking at a conference for eco-friendly start-ups or a national recycling convention. She could build a message that focuses on the environmental benefits of her innovation—a common link between the members of her audience.

## 4.6　Your Turn…

Now it is your turn to use this technique to define your own message. Take your time and think about it well. Test out your message on other people to see what they think. Is it clear? Can they remember it easily? We have provided you with the table overleaf to fill in your answers to the questions (Table 4.7).

**Table 4.7** Fill in this table to define your message

| | |
|---|---|
| WHAT IS YOUR OBJECTIVE? | MY OBJECTIVE IS... |
| WHAT ARE YOUR AUDIENCES' EXPECTATIONS? | AS [_____], WHAT I AM EXPECTING FROM THIS PRESENTATION IS... |
| WHAT ARGUMENTS DO YOU HAVE TO ACHIEVE YOUR OBJECTIVE FOR THIS SPECIFIC AUDIENCE? | MY MAIN ARGUMENTS ARE... |
| WHAT IS YOUR MESSAGE? | THE ONE THING THAT I WANT MY AUDIENCE TO REMEMBER FROM MY TALK IS... |

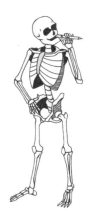

# 5

## Establish Your Plan
### Know Where You Are Going and How

We have all heard of stories from the past, where a handful of brave soldiers defeated armies in battle, greatly outnumbered in both size and force. During medieval times, at the request of their Kings and Queens, nations could spend years perfecting their defensive and offensive battle strategies. Essentially, their greatest weapon was solid preparation and robust tactics. When public speaking, your stage is your battlefield. You, standing up in front of a crowd of people, will undoubtedly be outnumbered by your audience: 1 to 10, 1 to 100, 1 to 1,000 or even more. To conquer them, you need a strategy. To succeed, you need to plan.

© Springer Nature Switzerland AG 2020
A. Youknovsky and J. Bowers, *SELL YOUR RESEARCH*,
https://doi.org/10.1007/978-3-030-34181-7_5

## 5.1     A Plan: The Basics

OK, so presenting your research is not exactly some blood-bath war from the Dark Ages, but there is something to be said about having a strategy. You need to know where your destination is and how to get there, especially if you only have a limited amount of time to do so. After defining the message you want your audience to take home with them, you need to conjure up a strategy that uses your content to defend it.

### 5.1.1     What is a Plan?

A plan is effectively the structure of your presentation. It provides you with a run through of the different elements you want to include in your talk and the most effective order in which to bring them to the table. Depending on your message the content and the structure of your talk will be different. Stay focused on your main objective. To successfully deliver your message to your audience, it needs to be at the forefront of your mind when producing a plan for your presentation.

### 5.1.2     Examples of Classical Structures

If you look at the literature, you will find numerous types of structures that can be used in presentations. In science, we see certain types more than others. For example, scientific subjects usually lend themselves well to a 'problem-solution' structure. Research tends to have a specific problem that is already laid out and the experimental results generally lead to either a solution, a proposed solution or a failure to provide a solution (which is nothing to be ashamed about!). We also often see a 'cause-effect' structure. For example, if you decide to take this approach, you may say:

> *We changed the fuel used in the rocket launcher engine from hydrogen to methane. With this $CH_4$ system, the tanks can be lighter and smaller, making the whole launch cheaper.*

> Cause (you changed the fuel) + effect (you will save money)

The 'compare-contrast' structure is also common in science: it allows you to compare different solutions with arguments and counterarguments, then balance them out to defend your conclusion. Finally, you could also opt for a chronological structure to provide a more time-specific list of facts or events.

It can be used to trace the path of a discovery from its early days until now or to give an historical account of a scientific idea or a technology.

### 5.1.3   When to Give Background Knowledge

When talking about any research topic, there is almost always a minimum level of context you need to provide. As we have already mentioned, you should be aware of the prior knowledge of your audience so that you can best adapt this part of your talk to their needs. To keep your audience engaged, you can play around with where you put background knowledge into your talk. And it does *not* always have to be at the beginning. It can be very nice to knit it in throughout the talk to avoid having a big block at the start. A bit of air between the real hard stuff can be handy! On the contrary, for very complex topics, don't be afraid to spend most of your talk on the background. Do what is needed to help you get your message through.

### 5.1.4   What Does a Plan Look Like?

A plan should help you define the different sections in your talk, with an overview of the inner structure of each of those sections. Your primary goal as a science communicator is to transmit your message and be understood. For that purpose, the construction of your argument should be logical. A typical plan will likely look something like this:

- Introduction,
- Section 1,
- Subsection 1.1
- Subsection 1.2
- Subsection 1.3

- Section 2,
- Subsection 2.1
- Subsection 2.2
- Subsection 2.3

- Section 3,
- Section …

- Conclusion.

Looking at the bullet points like this seems very basic but as already mentioned, clarity and simplicity are key. As soon as you put words to these sections instead of calling them 'Sects. 5.1 and 5.2', it will feel more natural.

However, whilst your plan may look like a linear sequence of parts, we believe that it is not the best way to prepare your structure. It is surprisingly limiting. For example, your thoughts go from top to bottom, which is not so helpful in indicating the relative importance of each idea or section. Also, if you go over one page, then you will be potentially breaking up your thought process. To overcome these limitations, we advise that you build the structure of your talk by using a mind map.

## 5.2    Mind Maps

Nowadays, most people have heard of the mind map, also referred to as a spider diagram. You may be surprised to discover that there is historical evidence for mind maps dating back as far as 3 BC, used by philosophers such as Porphyry of Tyros. However, they really found their revival in the 1970s thanks to a TV host called Tony Buzan, who introduced the idea of semantic networks under the name 'mind map'. In any case, it is a fantastic tool for getting ideas out of your head and onto the page in a way that helps you organise them logically. Mind maps can be created with nothing more than a pen and paper or, if you're more of a tech-creative then mind mapping software exists.

### 5.2.1    How to Use a Mind Map

We train speakers to use the tried-and-tested mind map method to plan their presentations. Figure 5.1 shows a typical structure. You simply write your message in the centre of your page and draw a circle around it. From that centre point, you can branch out each of your sections. The first idea, or the introduction, is added in the top right-hand corner of the page. Then, each of the following sections are added in a clockwise fashion, arranged around the outside of the message in the centre. Each of these main sections can branch out to other sub-sections and then they can do the same, and so forth.

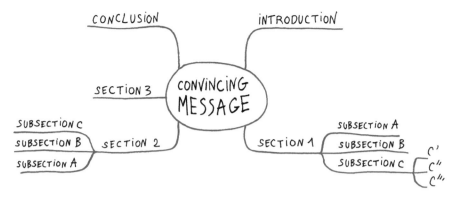

**Fig. 5.1** A mind map for planning a presentation. ©Agent Majeur

### 5.2.1.1 Message

Your message has its place right in the centre of the page where it can always be seen. Every argument in your talk should be related to your message in some way, and therefore each of your main ideas will branch off it. You should define your sections and subsections with key words, not full sentences, in order to keep your map easy to read.

### 5.2.1.2 Introduction and Conclusion

You need to think about how to start and end your talk. However, during the planning stage, we suggest that you leave these to one side for the moment. They involve creating a successful hook which we will discuss in Chap. 6. You will add the introduction and conclusion into your mind map later.

### 5.2.1.3 Sections

At this stage, you need to divide your talk into sections. The number and length of these parts will vary. We usually recommend that you have some-where between two to five sections in a talk, with three being the optimal. Keep in mind that if you have too many sections, your audience might get lost in your explanations.

### 5.2.1.4   Subsections

Each section can be split into smaller subsections. These are different points that can be discussed for each of the main ideas. Each subsection can then be divided into sub-subsections and so on. Also, watch out as the subsections should be arranged in a clockwise fashion. Hence, on the left-hand side of your mind map, the order of your keywords will be from bottom-to-top, as opposed to top-to-bottom on the right-hand side.

## 5.2.2   Benefits of Mind Mapping

The mind map is a simple and secure tool which allows you to make a plan in a visual way. Everything you write will fit on one page and can therefore be observed at a glance. This means that you will not get lost flicking back and forth between the pages in your notes. It also gives you a complete vision of all your ideas with key words, links between them and their relative importance to one another. Observing your thoughts in this way is particularly helpful for memorisation.

Putting ideas into a mind map format also favours creativity because in just one look, you may find other links or associations between concepts that you had never seen before. Similarly, the structure facilitates the addition of new ideas whilst avoiding corrections. Each new contribution can be organised within the schematic by simply drawing a line in between two others, so you can avoid scribbling out bits on your page.

## 5.2.3   Putting It into Action

Earlier in the chapter, we outlined some of the more common structures you may choose to follow. Whilst you can bear them in mind, we suggest that you let the order of your presentation happen naturally. We advise that you devise a plan that seems logical to you, but do not be afraid to play around with it. What happens if you give your major result first? Can you keep your audience curious by saving your most surprising video until the end, or should you show it right from the start? The best plan is the one that serves your message while keeping your audience engaged.

In the next section, we will go through some examples of different plans and how/when to use them. We will focus on the different sections of your talk without the introduction and conclusion, which will be built in later.

## 5.3    Case Study IV: The Technological Innovation

There is a standard, widespread presentation format within the scientific community that many researchers seem to use. Here are the different sections:

- Background
- What did we do?
- What was the result?

This structure does get the message across, but it is not always the best way to make a strong argument. Let's look at an example. An engineer has been working on a new factory process with the aim of improving productivity of a product line: plastic tube joints. He has been invited to speak to the board of directors for the factory about the changes he suggests and the impact on overall production. He knows his audience is a mixture of fellow engineers, quality control managers, the factory head and the financial director, so both specialists and non-specialists. Using the four-question process we presented in Chap. 4, he has come up with the following message:

> *You will increase productivity by 10% per year if you implement our new, improved factory process.*

He knows he has three main points to address because there are three steps in the factory process that he suggests changing, each resulting in an increase in overall production yield. The three features of the procedure that he suggests changing are:

- Mixing of starting materials,
- Washing of extrusion heads,
- Packaging of end products.

As you can see in his mind map in Fig. 5.2, he has decided to include three main sections in his presentation:

1. Background information
2. Changes made
3. Benefits on productivity.

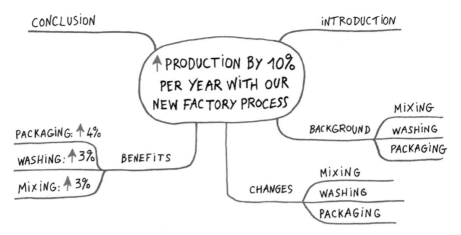

**Fig. 5.2** An example of a mind map for a classical plan. ©Agent Majeur

In his background, he introduces the three steps in the factory process that he will be talking about in his presentation: mixing, washing and packaging. Here, he plans to describe how they work and what's wrong with them. Then, when he describes the changes made, he refers to each of the three steps again: mixing, washing and packaging. Finally, when he talks about the results obtained and therefore how they will lead to a 10% increase in productivity, he mentions the benefits obtained for each, which are increases of 3%, 3% and 4% respectively.

This sequential description of the engineer's research certainly gives us all the relevant information in a logical manner. However, there is a more engaging way to present his topic. In our second mind map (Fig. 5.3), we have taken the same elements and arranged them differently. We suggest putting the steps of the process—mixing, washing and packaging—as the main sections. As subsections, we propose describing the benefits on productivity first, then the background and changes made of each step individually.

So, why do we suggest such a change? In the first example, the engineer is keeping all the good stuff until the end, which is much less exciting than putting it as early as possible. In particular, the financial director and head of the factory in his meeting want to know the benefits of the changes that he is suggesting. He will highlight the gains obtained by mentioning them first. Straight away he can say, "By changing the way you carry out mixing, you will improve your productivity by 3%". Now, his audience is interested. Now, they want to know how he got that result. They will be much more curious about the other details than if he starts by explaining the background

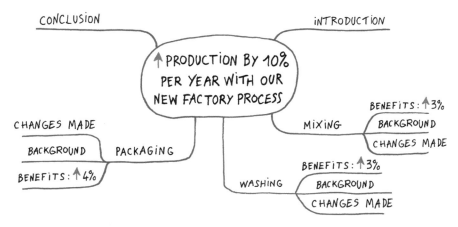

**Fig. 5.3** An example of a mind map for a punchy plan. ©Agent Majeur

to a potentially complex or tedious engineering procedure in the first few minutes.

The information included in his talk is the same, only the structure has changed. Plans like this are useful when you present a topic which is not too difficult for your audience to understand. Clarity being paramount in a scientific presentation, start with background information when needed, but dare to structure your argument differently whenever possible.

## 5.4 Case Study V: The Success Story

In the following example, we will demonstrate the problem-solution structure using a story of a postdoctoral biologist we worked with. She was developing a new vaccine for seafood allergies. She made a pitch to the scientific board who had already selected her as a grantee for postdoctoral research funding.

In the below mind map (Fig. 5.4) you will see the structure of her talk. As her first section, she chose to present the problem as the seafood allergy. In the second, she proposed a solution: a vaccine.

Looking at the detail of her plan, we see that for the 'problem' section she explained that three million people in the EU suffer from a seafood allergy—perhaps more than one would think! Then she described the nasty symptoms and, finally, that there was no cure. Clear and simple, the problem was posed. She then turned to her next section, which was the solution. Whilst this section was more specialist, she was able to succinctly explain that by introducing seafood DNA to the immune system, a higher tolerance to seafood

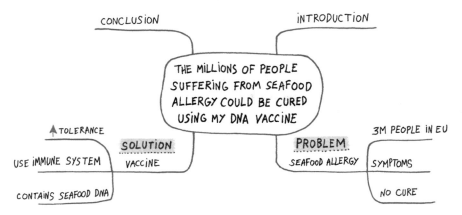

**Fig. 5.4** An example of a mind map for a problem-solution structure. ©Agent Majeur

could be stimulated in affected people and that this could be done with a vaccine. *Voila!* Problem-solution.

The simplicity of this plan is the strength of its success. Her pitch was a mere 3 min, but not one person in the room left without understanding her work. On the scientific board were a mixture of experts and non-specialists in her field. The plan was well adapted to their needs. The biologist had not yet completed the work as the project was in its early days, but she was able to present the utility of her work, reassuring the scientific board in front of her. Do not be afraid to go for a simple plan, especially for a short talk. The easier it is to follow, the clearer you will be to your audience.

## 5.5    Case Study VI: The Controversial Topic

One of the great things about science is that it is never set in stone. Our knowledge of the universe is constantly evolving, and new data can provide information that we never knew before. Unfortunately, because of this uncertainty, not everyone agrees on the same interpretation of information. This can lead to a topic being up for debate or even becoming controversial.

When giving a science talk, you may come up against an audience whose prior opinion of your topic goes against your own. This can be anything from physicists who prefer one theory of the universe over another, right up to large groups of people within the general public campaigning together against vaccines.

If you know that your audience does not share your interpretation of the data, then consider your topic as controversial. When faced with such a situation, we advise using a 'compare-contrast' structure for your presentation. In this type of plan, you are giving one point of view and then comparing it to another, in order to provide an argument for your message.

Our example here is from a Ph.D. Researcher in agronomy. She was presenting her research to an audience of various people from the French National Train Service. She was studying an alternative to using toxic pesticides to kill plants growing on railway tracks all across the country. The common stance in the rail service was that plants on the track were a menace and hazardous, so must be removed at all costs. Her research was going against the grain, so to speak, and she knew that most of the people in her audience would not agree with her.

The message of our agronomist was this:

*To avoid using toxic pesticides on railway tracks, we can grow plants on them.*

Figure 5.5 is the plan for her talk. In the first section, 'compare', she reminds the public that plants are forbidden on railways. There are internal regulations in place against plants on train tracks and she provides the reasoning behind them. She defines the potential risks, such as trip hazards on the platform and track. She also adds that to remove them, we currently require pesticides like glyphosate which can be toxic and will soon be banned from the market. Therefore, another solution is required.

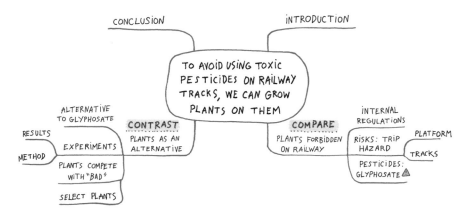

**Fig. 5.5** An example of a mind map for a compare-contrast structure. ©Agent Majeur

In her second section, she presents her contrasting argument: plants can be used as an alternative. She describes how to select species that are non-problematic and how they could be used to compete with unwanted vegetation. She then goes on to discuss the experiments she was carrying out on service tracks and the positive results she has obtained. As a final point, linking her 'contrast' section to her 'compare' section, she points out that her research suggests plants could offer a suitable alternative to glyphosate use.

We suggest this structure for controversial topics because it starts with what the audience already believes. You will never convince someone of something new if you come in all guns blazing from the outset. You need to demonstrate to your audience that you understand their opinions and why they may think that way. Creating this connection is vital if you are going to have a chance at convincing them of another point of view.

## 5.6    Before We Move on…

We suggest that now is a good point for you to stop and have a go at making your own mind map. Take a clean sheet of plain paper and put your thoughts down onto the page. Keep your message in the middle and think about the most effective way of convincing your audience of its significance.

# Part II

## Second Step: Make Your Envelope

# 6

## Hook Your Audience
### Catch Their Attention Right from the Start

Just because you are standing at the front of a room full of people doesn't mean they are actually listening to you. They may be looking right at you even though their mind has completely wandered astray, thinking about their own problems or the tasks they have to get done when the conference is over. As a speaker, it is your role to keep your audience engaged. When giving a live talk, the focus should be on catching and maintaining their attention. They want a reason to listen to you. Therefore, you need to reel them in with an idea that piques their interest or appeals to their curiosity. That is what we call a hook.

Crafting a hook is the first thing to do in the second stage of the SELL Method. You may not know exactly what your hook will be until later in the preparation process, but it is worth keeping this in mind whilst the cogs of your brain are whirring away. The more appropriate and original the idea, the more impact it will have.

© Springer Nature Switzerland AG 2020
A. Youknovsky and J. Bowers, *SELL YOUR RESEARCH*,
https://doi.org/10.1007/978-3-030-34181-7_6

## 6.1    Why Hook Your Listeners?

Your audience's attention is at its peak at the beginning of a talk. Then, over the course of a presentation, it usually wanes, with a rise again towards the end. This demonstrates how easily people can switch off as your talk progresses. In order to capture and maintain their attention, you should use a hook. Not only can a hook appeal to the curiosity of your audience, it can also:

- establish a sense of relevance to their own lives,
- present the global importance of your talk,
- surprise them,
- wake up their imagination,
- demonstrate your originality.

In other words, it's a great way to get them involved, making a connection.

## 6.2    When to Hook Your Listeners

Your optimal moments for using a hook are during your introduction and conclusion when your audience are most likely to be engaged. Whilst not obligatory, returning to your initial hook at other points in your talk can be a good way to pull your audience back in too. Therefore, the techniques we describe in this section can be used throughout your presentation, in order to maintain your audience's attention and keep them on their toes. In your introduction, you want to use your hook so that they will listen. During the conclusion, your hook or a memorable moment will ensure that they remember you. We can therefore add details into your mind map (Fig. 6.1).

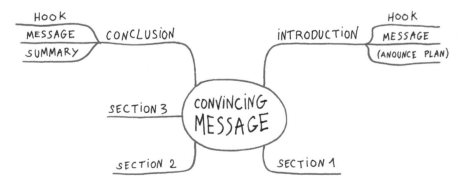

**Fig. 6.1** A mind map including details of the introduction and conclusion. ©Agent Majeur

## 6.2.1    Introduction

You should now be on your way to beginning to write your introduction and conclusion. These parts will sandwich the rest of the sections in your plan. For your introduction we suggest you include:

- your hook,
- your message.

You may also outline your plan, although it is not compulsory, especially for short talks. A science talk is not a murder mystery; you do not have to wait until the end of your presentation to reveal the message. You should bring it in as early as possible, ideally in your introduction, as it explains why your audience should bother listening. This is only possible, of course, when talking about subjects they can understand easily without long explanations.

If you want our top tip about opening lines, steer well clear of that over-used opener, "Thanks for inviting me, bla bla bla…". Whilst it is courteous to the organisers, such formalities send a signal to the room that your talk lacks originality and will be as mundane as every other presentation. Stay unique. Do something different that will surprise your spectators. Originality doesn't mean forgetting to be polite. You can always leave your thanks until the end of the introduction, or even the end of the presentation.

## 6.2.2    Conclusion

When getting to the end of your talk, you need to repeat all of the important information. Remember our Churchill quote: use a pile driver. That is the reason why your conclusion should:

- use a transition word to signal the end is near,
- clearly summarise your key points,
- remind the audience of the message,
- use a hook: the same as in your introduction, to connect the beginning and end of your talk, or another one.

A transition descriptor—'in conclusion', 'to summarise', 'to recap'—will allow your audience to clearly understand where you are currently up to in your talk. We advise that you summarise it too. There are two reasons for this: firstly, there is a possibility that at least some of your audience have not listened all the way through your presentation. But we hope not! Secondly, even if your audience has listened to you with great attention, they may not have held everything you have said in their mind. You need to point out the information that is important to remember. A good piece of communication relies on repetition.

We suggest also incorporating a memorable moment where possible at the end. This can be a return to your hook, maybe something funny or unexpected. But it can also be completely new, like a closing demo or visual that completes the loop of your talk.

## 6.3    Hook, Line and Sinker

To be effective, a hook should be more than just surprising or amusing. It should serve the purpose of delivering your message. The introduction is a two-in-one formula: 'I grab the attention of my audience and I get straight to the point.' Science talks do not have to be boring, but they do have to be relevant to the subject at hand. Here is a list of some of the most common hooks you can use:

- quote,
- joke,
- anecdote,
- analogy,
- question or challenge,
- picture, drawing, photo, or video,
- piece of music,
- link with current affairs, historical facts, or events,
- sample,
- demonstration.

### 6.3.1    Quote

Probably one of the most classic hooks is to use someone else's catchphrase. Let's face it, most of us are not Marie Curie or Oscar Wilde, so there is no harm in accepting that these articulate well-known figures have already managed to say what you want to say, only much better. There are several reasons why you may want to use a quote from someone else—famous or not. Here we list some examples.

To make people smile:

> *Gravity is not responsible for people falling in love.*
> Albert Einstein, Physicist

To introduce unexpected results:

> *To know the history of science is to recognise the mortality of any claim to universal truth.*
> Evelyn Fox Keller, Physicist

To promote interdisciplinarity:

> *This is an era of specialists, each of whom sees his own problem and is unaware of or intolerant of the larger frame into which it fits.*
> Rachel Carson, Marine Biologist

To indicate the importance of your field:

> *There's one issue that will define the contours of this century more dramatically than any other, and that is the urgent threat of a changing climate.*
> Barack Obama, US President

There are many quotes circulating the web, some of which are false. Whilst it's almost impossible to truly verify very old quotes, do your best to find a reliable source. Please, also be conscious of who you are quoting and why. Far too often, speakers quote middle-aged, white men because that's what history has provided. Think about quoting women or people from minority groups to avoid looking lazy. And don't borrow quotes that are over-used. Phrases like "Size does matter," or "With great power comes great responsibility," are used again and again in science.

## 6.3.2  Joke

Humour is one of the best tools you can have up your sleeve. A good joke is enough to get anyone on your side and lightens the mood of your audience. When talking about using humour in a talk, public speaking expert Tom Antion says that it is an "attention-gaining device" that should be spread out through the presentation and not used all at once. Making someone laugh, or at least smile, and feel positive is a great way to pique their interest. Like all good storytellers, when making a joke you must commit to it. A half-hearted delivery won't do you any favours, nor will telling a joke that is likely to offend anyone.

Humour is one thing but bear in mind that no one is going to confuse you with a comedian—keep it clean. Don't make gags about gender, race, disability or any other sensitive topics. You will sound offensive, even if it's not your intention.

If you do decide to use a joke as your hook, remember timing is of the utmost importance. You should know exactly when and how you will deliver it. Test it out on someone beforehand, if possible. Like any good presentation, a well-practised joke looks the most natural. Also, don't use humour if you're not comfortable with it. Authenticity is the key to a great presentation, so don't pretend to be someone else.

Jokes can be surprisingly simple. One of our Ph.D. trainees who entered a pitch competition started his talk with, "I study the science of failure," which in itself is not necessarily funny. However, his thumbs-up and big, dorky smile added just the right dose of comedy value to get a laugh from the room.

### 6.3.3 Anecdote

An anecdote is something that occurred to you or somebody close to you; a story that you are going to share with your audience. It can be related to a trip that you went on, a person you have encountered, your job, your family, or even your dog—as long as it has something to do with the message you want to get across, anything goes.

Anecdotes are good for two reasons: they tell a real story, contributing to your credibility with first-hand evidence, and they help you to appear more human by giving details about yourself. They give you a chance to tell your audience a personal tale: "When I was seven years old, my parents bought me a brand-new bike. At the time, I felt like it was the best thing that had ever happened to me. Until I got my PhD!" When you share a personal story, which is genuine, it lets your audience into your life, showing who you really are.

The important thing to bear in mind is that its purpose as a hook is to help you make a connection with the audience. You can use an anecdote to create a link with them by finding something you may have in common, or a funny detail about yourself. One of us worked in a science museum and regularly presented a workshop to primary school-age children. The demo was easy enough: feeding frogs in front of youngsters whilst explaining how they differ from toads and the senses they use to locate prey in the wild. Simple. Except the presentation was in French and the presenter was a native

English speaker who still, even now, cannot pronounce the word *grenouille* (frog, in French). He started his demo with, "Normally when I have to say frog (*grenouille*), I just say toad (*crapaud*) instead because it's easier to pronounce. Unfortunately, today I can't do that as I'm here to talk to you about the difference between the two! So, you'll have to forgive me for the way I say gr-nou-ille." It worked well, as it was a true story and it showed vulnerability whilst making the kids laugh, helping to catch their attention.

Anecdotes don't have to be funny. As the hook to his acceptance presentation for a science prize, one of our trainees Dr. Salvatore Spicuglia used the following hook:

> *Chance plays an important role in scientific research. As a young scientist, I was profoundly moved by a book written by François Jacob, winner of the Nobel prize for medicine in 1965. Thanks to him, I was able to understand that research is above all else a vocation of endless wonderment, passion and an openness to be able to welcome chance. Throughout my career, chance has been important most notably in my most recent discoveries in epigenetics.*

This anecdote is more serious than funny. He is providing us with a glimpse into himself as a real human being, something that we can engage with. It gives us a look into the life and mind of the speaker, helping the audience to truly connect with him.

## 6.3.4  Analogy

In science talks, analogies are often used to describe a complicated concept using a more commonly understood one. This works particularly well as a hook for general public audiences. For example, a classical analogy a geneticist may use is:

> *Inside every cell in your body there is DNA, a sequence of code. If we consider that code to be the letters of the alphabet, then the DNA is like a book. It's my job to learn how to read that book.*

The idea here is to describe DNA as a book. Whilst it does not resemble a book in any shape or form, the analogy helps to describe its purpose and the role of the geneticist. But be careful, some figurative expressions like this one are over-used. Try to be original.

That's what another speaker did with the introduction, "Until now, you have always listened with your ears and seen with your eyes. What if you could listen with your eyes?" She was in fact presenting an innovation that collects acoustic information using an optical system. The technology she was referring to would seem to be counter-intuitive, hence why she highlighted in early on. With a simple analogy, she was able to make her rather complex technology understandable to all. It was intriguing for the audience, as they asked themselves how such a device could possibly work.

## 6.3.5  Question or Challenge

A great way to reel your audience in is to push them to interact with you. Make them feel part of your talk. Something as simple as asking a question can do this. A trainee of ours once used the following question to hook his audience: "Let's all imagine that we are out in the sun. What happens to your body?" It was a great way of making the entire audience feel involved. Not only were they required to use their minds to create an image of being out in the sun, it also made the talk relevant to the whole audience—after all, who *hasn't* been out in the sun before? He managed to guide the spectators from the idea of the sun penetrating the skin, to using laser therapy to attack tumours in cancer patients.

Setting a challenge to the audience, or to yourself, can be a powerful tool. At a recent conference, one speaker took to the stage to talk about personal data security on smartphones. As soon as his talk began, he asked everyone in the audience to take out their devices, unlock the screen and open their last text message conversation. He then said, "Now hand your phone to the person sitting next to you." The theatre was filled instantly with hundreds of awkward mumbles, as most people felt very uncomfortable doing as asked, and subsequently a vast majority refused to give their device to their neighbours. The point he was making was that if you do not even like the idea of the person sat next to you looking at your private data, then why give companies access to it?

When asking a question, be sure to know whether you want a response or not. If your question is rhetorical, it is there to make your audience think. We advise that after posing a question, you leave at least two or three seconds of silence. Even if the audience is not meant to answer out loud, they need time to imagine the answer in their head.

One of our trainees who won a pitch competition, started his talk with a question: "What is the most common surgical procedure carried out in France?". Instinctively, what would you reply? Appendicitis? Wisdom teeth? The correct answer is, in fact, the cataract. There you go. Very quickly, the speaker managed to make us understand the importance of his topic; an innovation that will improve cataract operations carried out over a million times each year in France.

Last but not least, don't ask irrelevant questions to which everybody knows the answer, unless you plan to provide an unexpected one. For example, "Who here has ever arrived late to a meeting?". In this case, it's preferable to switch it to an affirmation, "We have all arrived late to a meeting at some point in our lives".

### 6.3.6 Picture, Drawing or Photo

One decent, thought-provoking picture can be all it takes to intrigue a room full of people. Imagery can also evoke a strong emotional response in your audience: anger, surprise, laughter and so forth. We trained a sociologist who was studying the effect of hierarchy on decision-making. At the beginning of her presentation she presented a photograph of two young girls laughing, which she flashed on the screen behind her. They looked happy, excited and youthful. She stated their names, then paused before saying, "They were both killed in an aeroplane crash". The room fell quiet. Whilst it is a sombre topic, the way that the speaker used the image certainly caught everyone's attention. The spectators immediately grasped the importance of what she was about to tell them. The subject of her talk was her study of the dangers of hierarchy between captains and their co-pilots in aeroplane cockpits. Had she only referred to pilots, we may not have felt so concerned. Whereas, raising the issue of danger to passengers in her hook, the audience were more readily emotionally connected.

On another occasion, one of our trainees produced a striking visual to complement the hook for his talk. He was carrying out his research on the economic viability of sustainable energy sources. He started his talk by saying, "I work in economics, but my topic is actually more like the job of a dietician." His visual was of a person sat at a dinner table with a meal, using a salt and pepper shaker in the form of a solar panel and a wind turbine. His analogy compared finding the right balance in your diet with finding the right balance of energy sources—coal, nuclear, solar or others—on the scale of a country. The visual worked very well because it was intriguing, but also summarised his research field nicely.

### 6.3.7 Video

If you want your presentation to be more dynamic, you can include video clips or short animations. Used as a hook, a video shouldn't last more than one or two minutes. You don't want your whole talk to be your audience watching a film, but it can help to make a good point. It is a great way to show your audience what they cannot see there and then in your presentation room. You can use a video you made yourself in your lab, or a clip of someone relevant to your message who cannot be there in person at your talk.

Let's take the example of a scientist studying chimpanzee communication. She and her colleagues decode the meaning of hand gestures used by wild primates to communicate with one another in a very basic form of sign language. When presenting her research findings to a TV studio audience, she included a video clip of two chimpanzees communicating via hand gestures. She first showed a clip of one signalling to another. The image was cut just before the second chimpanzee responded and she then asked a question to the amphitheatre: "What do you think this gesture could mean?" After they guessed, the researcher went back to the clip to show the second chimpanzee approach the first, proving her point that the original hand signal meant "come here" in chimp!

Be careful with the quality of the videos you choose. Both the image and sound should be good enough for your intended use. For a large room, with a big screen and a powerful sound system, the quality needs to be better than for a regular small meeting room.

## 6.3.8   Piece of Music

Using music as a hook is rarer but can be very effective in certain situations. A very common sound that everyone knows, like a famous phone ringtone, the jingle that comes before a train announcement, or a theme tune for a TV show, can effectively provide a point of reference. Music can evoke different types of emotion, like love, sadness, or joy, and you can also use it to add humour.

The principle constraint with music is copyright. You must make sure that your music is royalty free or that you have the right to use it. This is also true for videos and photos. One way of getting around this is to sing or hum the song yourself, as any 'live' reproduction of music is legally allowed. However, this is only appropriate in some circumstances. After all, this is a science talk, not the X Factor boot camp.

### 6.3.9    Link with Current or Historical Facts

By referring to an event in the news or a historical fact that everyone is aware of, you are calling upon your audience's own experience of the world around them. For example, you can rely on a piece of current affairs as a hook:

> *For several years now, food poisoning has become common in China. Last month, it was the turn of schoolchildren to be affected. One even died, whilst 31 others were taken to the nearest children's hospital.*

Historical facts can also work in a similar way, providing a gateway for your audience to create a connection between what they already know and what you are about to tell them. An example of a good use of imagery was in a talk about wireless technology. The speaker used an old black and white image of a telegraph room on a ship. He asked the audience if they recognised it. They all shook their heads. He then explained that it was, in fact, the radio room of the Titanic. He added that it was equipped with a type of wireless technology that may now be used for space launchers. Whilst the use of Titanic may not be a good example for the safety aspects surrounding his technology, it was a good point of reference in terms of its age. Most people will at least be able to estimate the sinking of Titanic to have occurred at some point in the early 1900s, which makes the technology—surprisingly—well over 100 years old.

As with jokes, it is worth being considerate about the examples you use. The choice of historical facts as a hook may put the speaker in a difficult situation in certain circumstances. History and society are deeply linked. Some people may have different beliefs about historical or societal 'fact'. Again, knowing your audience will help with this.

### 6.3.10    Sample

Unfortunately, you cannot always invite your whole audience to your lab with you. But you can reveal a piece of your research to your spectators in the form of a sample. This could be a new mechanical valve you've been working on, or perhaps a particularly weightless floor tile you designed. Samples are both efficient and concrete. It's a special type of interaction, because the audience can be invited to participate with all senses: touch, sight, smell, sound or even taste. It can also help to prove a point if your message involves the development of a new product.

This method is at its optimal with small groups. Be aware that you will need to pause during your talk whilst the audience is passing the sample around, because they will be much less attentive to what you are saying if they are fiddling with an object at the same time. If you start handing out samples in an auditorium, it is likely that by the time the people sat up at the back have it in their hands, you will have moved on in your talk—unless you have enough samples to hand out to everyone. But then you may break your flow if you must stop and hand out a hundred test tubes. If you are giving out a lecture theatre-worth of samples, you can provide them beforehand if you trust your audience won't be distracted. Otherwise, you can hide them underneath the seats or save them until the end of the presentation. In scenarios where you cannot hand out a sample to everyone and your audience may not be able to see, you could use a live camera hooked up to a projector.

Beware, if your samples are fragile, make sure you communicate that clearly to the audience, so they know to handle with care. You certainly don't want someone to drop your €7 million microchip behind the radiator, never to be found again!

### 6.3.11  Demonstration

Finally, the most spectacular of all supporting materials is a live demonstration, and we will come back to this in Chap. 8. Any talk will benefit from you showing something in action. There's nothing better for proving your point to your spectators. Of course, the more remarkable, the bigger the impact. However, a demonstration does not necessarily have to be a huge explosion with pyrotechnics and glitter. It could be a scale model of a design, a piece of equipment or a demo that represents a difficult concept for your audience to grasp with words and images alone.

The benefits of live demos are clear; they are captivating and add depth to your talk. Take care, however, to make sure that the demo works properly. Practise well in advance and try it at least once in your presentation room too, on the day if possible. Make sure you have everything you need. If your demo fails, it can break your confidence or lose your audience. But, if it works, the payoff is worth it!

## 6.4  Knitting Your Hook into Your Talk

Bear in mind that your hook can be more than one of these things. It will often take a form that incorporates several, this is what we suggest for high-impact. Below is an example of a well-used hook, which fits into several of the above categories.

A speaker was presenting a new software development to the owners of a building company. He believed that the innovation was so important to the world of construction that, in the near future, it would become the computer model that *everybody* in engineering would use. Early in the presentation, as a hook, the speaker flashed up an image of a non-descript roll of sticky tape on the presentation screen. He then asked the audience what the name of the object in the picture was.

The presentation was in France, so the audience replied "Scotch." Had he been in the UK, the response would no doubt have been "Sellotape." However, it was neither. With a smile, the speaker informed the meeting room that the image represented was nothing more than "Sticky tape." Scotch and Sellotape are simply trademarks. They are brands that are so synonymous with sticky tape that we use the brand name when referring to the product. The hook worked and the spectators were intrigued. He re-used his hook at the end of his talk too. After he finished, one of his spectators said, "We want to use Scotch." Proof that his hook had worked!

In that instance, the hook took several forms. The speaker used an *image* of non-branded sticky tape. He asked the audience a *question*, one which was intended for them to verbally respond to as opposed to a rhetorical question. He offered the Scotch sticky tape reference as an *analogy* of the credibility of his innovation. Finally, the example also used *humour* because there was, of course, a laugh in the room when the audience all shouted out "Scotch!" and the speaker said "No! Sticky tape!".

## 6.5    Be Comfortable with Your Hook…

Bear in mind the importance of being at ease with your hook. If it's a joke but you only take it halfway, there is less chance that your audience will find it funny. If it's a demonstration, make sure you know how everything works and be ready for if it goes wrong. Remember, the hook sets the tone for your talk. When it suits you and your personality, it is a great opportunity to be original and make a good first impression.

# 7

## Design Your Slides

### Use PowerPoint to the Advantage of Your Talk

The scientific community seems to have adopted a long-held myth that PowerPoint slides are obligatory when giving a presentation. Bill Gates might not want to hear this, but yes, it is nothing more than a myth. Most of the world's greatest speakers, Martin Luther King, Oprah Winfrey, or even Plato certainly never relied on a set of wordy slides to support their arguments.

What is true, however, is that in science talks you will often have pieces of evidence to show your audience—results, tools, equations, that sort of thing. Therefore, your supporting presentation serves as material for your speech and can support your story in the form of data plots, graphs or pictures. They can also accompany the narrative of your presentation by providing depth and colour.

We know that people remember information better if they see and hear it together, rather than seeing or hearing alone. However, you should not rely

© Springer Nature Switzerland AG 2020
A. Youknovsky and J. Bowers, *SELL YOUR RESEARCH*,
https://doi.org/10.1007/978-3-030-34181-7_7

on your visual aids and they shouldn't replace you—slides with tonnes of text that you are simply saying out loud may as well have been sent in an e-mail, we don't need you to read them. The focus of your audience's attention is on *you*. Don't let your slides steal the limelight. As a side note, we refer here to PowerPoint, however, the same rules apply to other presentation software such as Google Slides and Keynote for Mac.

## 7.1    What's the (Power) Point?

Remember: what is the objective of your presentation? To deliver your message to your audience in the clearest way possible. In a presentation, it is you who is the transmitter, not your slides. The main challenge during your talk is maintaining the attention of your audience. You need them to focus on what you are saying. Presentation screens can very easily become a distraction. If your audience is caught up in reading or trying to understand them, they aren't paying you their full attention, if any at all!

All eyes and ears should be on you. Prepare slides that are as clear and concise as possible. When you start designing your PowerPoint presentation, we recommend you use paper far from your computer to give your creativity free reign. Draw your sequence in order of appearance, as if it were a movie storyboard. There is a plethora of literature out there about this topic, so we have concentrated on two main aspects that, as scientists, you should bear in mind:

- your slides should never take your audience's attention away from you, so you should make a **point with power**,
- you should apply the same aesthetic guidelines as creative professionals to achieve attractive slides by following **graphic design rules**.

## 7.2    A Point with Power

PowerPoint presentations are there to strengthen your argument. Far too often, we see scientists with overloaded slides, lost for breath whilst trying to comment on the different aspects of an extremely complex data plot. If you've ever tried to explain an intricate statistical image like a Principal Component Analysis to an audience, then you will surely know how that feels. Other speakers show images or text that they don't use. It is very difficult to understand their connection to the point at hand when no explanation or commentary is given.

Our take-home message here is that slides are distracting to your audience. They will look at them and try to make sense of what they see until they can understand. During that period of time, they won't be able to listen to you. You need to avoid this happening at all costs. PowerPoint expert Nancy Duarte invented the glance test, whereby anybody looking at your slide should understand what it means in under three seconds. It should be that simple. If anybody must think about its meaning, then they are not paying attention to you.

In this section, we provide you with several tips to help you create visual aids that can strengthen your talk, not weaken it.

### 7.2.1   Rules Are Meant to Be Broken

Another myth about presentation slides is that they should be a certain format. Scientists regularly seem to think that we should have a set number of words, a logo, bullet points, or images of a certain size, but really, it's up to you.

Slides are there to highlight specific elements of your presentation, so if you ensure good quality, concise content, you're on the right track. By differentiating yourself with your own personal style, you will highlight your topic and stand out from the crowd. *Be brave.*

### 7.2.2   Keep It Slick

PowerPoint slides are not meant to be self-explanatory. They are there to complement your speech. Use as little text as possible or no text at all. Remember—if your audience is reading your slides, they aren't listening to you. If you have complex information you really want to convey, think of including it on a paper handout which you can give out at the end of your presentation, or show a link to a website for more information. You can also rely on animations to piece a complex idea together sequentially, rather than giving it all at once.

If you choose to use a quote, make sure you find a short one. It is better to reuse it integrally in a slide to reinforce your argument. Again, if you match it with a visual aid, it will have more impact.

### 7.2.3 Use Readable Text

Sans-serif fonts are easier to read on a projector screen. Fonts like this include Arial, Helvetica, Tahoma or Univers. Try not to use too many words in capitals, as this will reduce readability. That rule also counts for font size: try to use a size that will be easy to read wherever people are sitting in the audience, which is usually a minimum of 24-point font.

### 7.2.4 Be Visual

Where possible, choose an image over text. It can often replace long sentences and will remain etched in your audience's memory. Make sure that it is immediately clear to the audience what the picture is. Unless, of course, you want them to think about it for a moment. In that case, stop and give them time to do so.

We once coached someone for a competition who was making a three-minute pitch about an interdisciplinary project. He used a well-known image that looks like either a rabbit or a duck depending on who is looking at it. It was actually the hook at the beginning of his talk, and he managed to really engage his audience by simply asking, "What do you see in this picture?" and waiting for a few moments for them to think about what they could see (Fig. 7.1).

**Fig. 7.1** In this image, originally published in *Fliegende Blätter* magazine in 1892, some people see a rabbit where others see a duck (creative commons license)

The speaker then declared, "You are all looking at the same picture, but you are not seeing the same thing. This is the subject of my thesis. I help people who see different solutions to the same problem depending on their expertise." An effective use of imagery to describe the bias that various specialists have when describing the same problem. We advise using good quality visuals, especially for photographs, that won't be pixelated when projected on a screen. Avoid cramming them into a small frame: give them the space they deserve.

## 7.2.5 Be Coherent

Try to maintain consistency between your speech and your slides. Do not paraphrase in writing the words you will say. If you describe something in your talk as loud, then you should write the word loud on your slides. Avoid using a different term like 'noisy' or 'high audio gain' as a replacement. Any inconsistencies like this mean that your audience will need time to assimilate the difference and work out that you are referring to the same thing. That could take seconds, but it could also take minutes in some cases. However obvious it is, if they are thinking about that, then they are not listening to you!

## 7.2.6 Simplify Your Graphs

A graph can be complex in a paper because people will sit down and take the time to go through it. During your presentation, they can't do that. You need to simplify your graphs to include only the elements you are referring to. If there are four bars on your bar chart but you only mention two of them, delete the two unnecessary bars. If you don't point them out, your audience will be distracted thinking about their relevance.

## 7.2.7 Use Animations

If you need to show a complex schema or graph, use animations available in PowerPoint to synchronise your explanation with your visuals. It is called 'directed attention', or the art of making your audience look where you want at a moment that you have decided. Your animations should be simple and not distract your audience too much. Simply making a visual element appear or disappear is generally sufficient.

## 7.2.8 Think About Your Transitions

The following slides are the most important, although none are completely obligatory. You should make sure that they are clear road marks in order to help your audience follow your presentation easily.

### 7.2.8.1 Title

This slide should announce the subject of your talk, so must have the title of your presentation along with your name and the organisation you represent. It can also contain an image related to your work, the date, your logo and perhaps the sponsors of your project. This will normally be projected before you begin.

### 7.2.8.2 Plan

A slide to indicate where you will be going with your talk. It can help you give a sense of what your audience can expect.

### 7.2.8.3 Transitions

These slides should help your audience situate themselves in your presentation. Examples include: "Part 1/2/3" or "Introduction, Methods, Conclusion."

### 7.2.8.4 Conclusion

You should announce here that you have come to the end of your presentation and give a quick refresher of your key points. Several slides can be used in a conclusion, so you don't have to cram everything into one.

## 7.2.9 Point Out Important Figures

Facts and figures can have a strong impact. But you need to drive home the important ones. If you have a data set or graph and you want people to pay particular attention to part of it, make it obvious: use an arrow, a clear colour-code or circle it.

Slides can be a great way to highlight important figures. For example, if you are communicating the fact that 20% of British people are obese or overweight, you could combine a picture of greasy fish and chips and a Union Jack flag with the 20% figure in bold characters. It may seem too obvious, even crude, but it can be understood straight away. Be careful about your choices and the connotations they may have, you don't want to offend your audience—likewise if it is supposed to be humorous, keep it clean.

### 7.2.10   Lighten Your Content

As a rule of thumb, it should be one idea per slide. Generally, you should favour slides with little text, in order to avoid diverting the gaze or mind of your audience. For the same purpose, your company's logo may appear on your title slide or at the end of your talk, but it is not necessary in between. Therefore, you may remove it.

Blank space is precious. It allows the viewer to relax and reflect on what they have seen. It helps to guide the eye through your slides and highlight important terms. Titles, for example, have more white space around them; they therefore have more importance in the document.

### 7.2.11   Choose Quality Images

If you use photographs, make sure they are the best quality photographs you can get your hands on. You don't want them to be pixelated when projected on a screen. Give them plenty of space too—why not even use a full slide for your picture? Another suggestion is to avoid using graphs that are 3D. For some reason, Microsoft Excel still offers this as an option. A 3D graph adds no extra logical dimension and makes it harder to read and understand—steer clear.

## 7.3   Graphic Design Rules

Whilst they are extremely creative people, graphic designers express their artistic ability whilst following a set of design rules. These help to create harmonious works of art that are easy on the eye and comprehensible to the viewer. Whilst it is a speciality of its own, as a scientist you can still take inspiration by learning and applying some of these techniques. They can help to create both aesthetically pleasing and easily understandable slides. Here are four rules used by graphic designers:

- contrast,
- repetition,
- alignment,
- proximity.

### 7.3.1   Contrast

To attract attention to something, it is best to use contrast. This can be contrast of the font, such as big letters, bold effects or colour to draw your audience's gaze to the place of your choice. Also, use a light-coloured text on a dark background, or a dark coloured text on a light background—easy, right? The contrast between the two is important.

### 7.3.2   Alignment

Correct alignment gives your slides structure and coherence. You should ensure your text and images are aligned throughout your document. This means having your titles in the same place on each of them, aligning your bullet points neatly with each other, and having your images the same size as one another, and so forth.

### 7.3.3   Proximity

Intellectual proximity is about putting the elements with the same sense side-by-side with one another. The connection between similar elements makes it easier to understand your slides. Content related to the same topic should be closer. With images, you can even sometimes overlap them, to suggest a link. Others you can separate onto different slides, or give them their own space, in order to indicate separate ideas.

### 7.3.4   Repetition

Try as much as possible to give standard unity to your materials. This means using a coherent colour scheme, sticking with the same font, and using similar style graphs where possible. However, you should only do this up to a certain point, because at the same time you should avoid monotony. Be coherent but mix it up with your own personal style.

## 7.4   Technical Glitches

Finally, once you have your slides ready and your speech is prepared, we strongly advise that you do a few additional things before you head off to your conference or meeting. When using PowerPoint, you will at some point, through no fault of your own, encounter technical problems. If it happens a minute before you present, it will do no good for your anxiety levels.

A common problem is that your slides are not compatible with the computer or projector in the room where you're giving your talk. Certain animations, videos or sound effects may not work. You may find that the font you used has somehow changed drastically when you open the file on a different computer. This can also involve the page format, such as whether they are in 16:9 or 4:3, for example.

Make sure you ask about the equipment that will be in the room or auditorium where you are presenting before you go and find out what file formats they require. If possible, take your own laptop that you can connect to the projector. Have a copy of your presentation on your computer or a USB key, with the PowerPoint file and also a PDF version, just in case. As soon as you can, test your PowerPoint slides on the projector in the room when you arrive. It will also help to put your mind at ease.

Finally, we advise that you use the 'presenter mode' on your presentation software. Most systems have this function, which allows you to see both the current slide and the next one on the list. It also means you can leave your laptop in front of you, so there's no excuse for you to have to turn to look at your projection. You should keep facing your audience whenever possible. Be aware that with some equipment, the presenter mode may not function as expected, so print a copy to help you memorise the sequence.

## 7.5    Slide Improvements

To understand more precisely how to apply these rules, we have developed an exercise. Here, we take an overly complex slide and deconstruct it, helping us to reflect on the content more easily. Take a good hard look at the sample below. It is an adaptation of a real presentation used by a scientist for a general public audience. For the rest of this chapter, we will be talking about this one slide. Yes, there is that much to say about it.

Before you turn over the page, we challenge you to analyse the sample slide in Fig. 7.2 and re-design it on your own. If you aren't near a computer, then take out a piece of paper and a pen and draw your answer. When you are ready, you will find our suggestions on the next page. To improve this example, we have used tools that you have access to on a daily basis. That is to say, visuals that you can make on your own using PowerPoint or buy for a few pennies online. You don't need to be a professional graphic designer to create attractive visual aids.

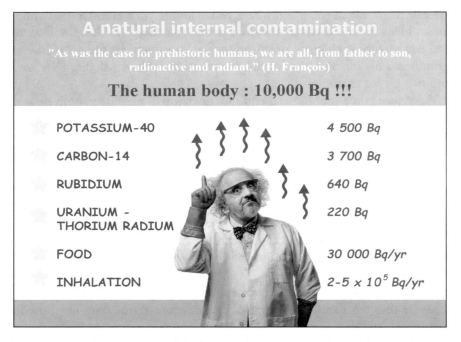

**Fig. 7.2** A sample PowerPoint slide from a science presentation with many improvements to be made

## 7.5.1 Initial Thoughts

Over the next page we suggest what we consider to be the best improvements this scientist could have made, to make a better point with power. But first, below are some of the design rules that this researcher has broken:

- Overall, there is **too much text**. Overleaf, in our suggestions, we have broken the slide into four to spread the information out, making it much more understandable.
- The choice of colours is not optimal. More than just a question of personal taste, the yellow title on a light blue background does not provide enough **contrast** to make it easy to read.
- There is a lack of **coherence** in both colours and fonts, resulting in an inharmonious appearance.
- Whilst the photo is funny, the slide is referring to the amount of Becquerels in any human body, not just a scientist. This picture should have been of a regular person for **clarity**.
- The image of the scientist separates the numbers from their corresponding text, making the connection between the two much **harder to understand**.
- Instead of bullet points, the creator has decided on large, yellow stars that **draw attention away** from the important information in the slide.

## 7.5.2 Give It Space, Let It Breathe

The sample slide has a quote at the top, but it is completely lost in an information overload. Not only is the text small and difficult to read, there is not enough time for the audience to take it in. Quotes are there to help get a point across. If you want your audience to understand its meaning and make the connection with your talk, then it can go on a slide on its own.

Our example of an improved first slide in Fig. 7.3 includes a simple, good quality visual that accompanies the message behind the quote. Give your audience all the help they need to absorb the point you are trying to make.

## 7.5.3 Don't Give Everything Away

On the sample slide, the speaker has decided to give all the information up front. This means that the audience can read ahead, and nothing is left to surprise. For example, it contains the sentence: "The human body: 10,000 Bq!" This statement is meant to be surprising for the audience, as it seems like a

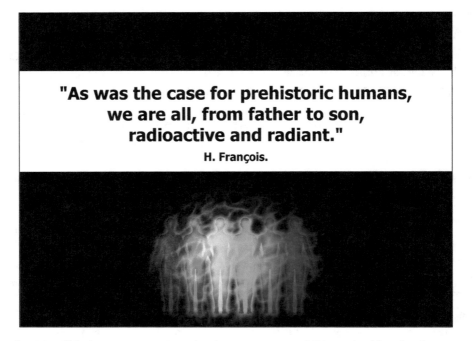

**Fig. 7.3** Slide improvement 1: make the quote more visible and add a visual

very high figure. We don't tend to think of the human body as having such high levels of radioactivity. But, by labelling it directly, all sense of surprise was lost.

In Fig. 7.4, we suggest taking that statement out and adding in a new one in the form of a question. The speaker can then ask the audience, "How many becquerels are there in the average human body?". He makes them think about it. Keeping your audience engaged and surprising them will give your stats more impact.

### 7.5.4    Help Your Data Speak for Themselves

Now that our presenter has asked the question; he can give his audience the answer: 10,000 Bq! The sample slide gives the different values for different types of radiation, but we suggest putting that data into a pie chart (Fig. 7.5). It's the best tool to show proportions, especially when the exact numbers are not helpful to an audience.

Or inversely, if you want your audience to have the exact figures, then a table will be more appropriate. Under the graph, for clarity, we have added a simple, easy-to-read legend.

**Fig. 7.4** Slide improvement 2: ask the audience a question to add an element of surprise

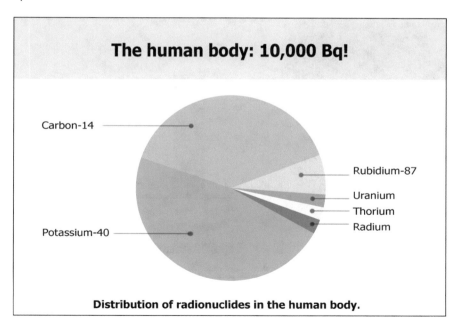

**Fig. 7.5** Slide improvement 3: use a pie chart rather than data to point out proportions

### 7.5.5    Point Out the Comparisons

You may not have even noticed, but at the bottom of the list on the sample slide there are two data points related to the origins of radioactivity in the human body. This means that two different types of information were listed together. Visually, it should be obvious that potassium, uranium, carbon and the other elements are not comparable to what the speaker is referring to as food and inhalation. The unit is not even the same: Becquerels (Bq) for the first, and Becquerels per year (Bq/yr) for the last ones.

A good option would be to put a final visual dedicated to radioactivity exposure, and again making a comparison between food and inhalation (Fig. 7.6). It is difficult to mentally compare the numbers 30,000 and 500,000 as they are not two proportions of the same whole, so a bar chart is more appropriate than a pie chart in this instance. We also propose a picture of a person with a flower to reinforce the message that inhalation is a more prominent source of radioactivity than food.

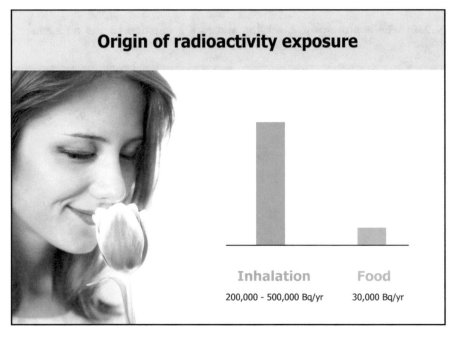

**Fig. 7.6**  Slide improvement 4: use simple graphs to visually compare data

## 7.6    In Conclusion…

More slides do not mean more speaking time is required. Far too often, scientists overload their visual aids with information. This is a big no-no! Take home message: one idea per slide and don't include irrelevant information. Your audience should be listening to you, not trying to work out what the content means. Apologies for repeating ourselves, but this is an important point. And you know how we love that pile driver Churchill talked about!

# 8

# Create Your (Additional) Supporting Materials
## Alternate Presentation Tools to Keep the Rhythm

By this point, your presentation should really be taking shape. You know your context, your message, you have a plan and now, your content is beginning to look vibrant. You should be well on your way to having a dynamic presentation with a good rhythm that suits your personality. Now is the time to take that a little further. Even using all the advice that we have provided you with, PowerPoint slides can nonetheless become monotonous. To improve the tempo of your talk, you can take your presentation out of the screen and into the room by making it interactive.

It is rare that you will turn up to give a presentation completely empty handed. Even for a short-form pitch competition, you may still have at least one slide, or an object as a prop with you. A dash of theatrics can work wonders for waking up sleepy spectators by alternating the supporting materials you use. It is worth noting here that some of them may be similar to what

© Springer Nature Switzerland AG 2020
A. Youknovsky and J. Bowers, *SELL YOUR RESEARCH*,
https://doi.org/10.1007/978-3-030-34181-7_8

we suggested as hooks. But why limit yourself to just the introduction and conclusion?

If you are lucky enough to take a part in a presentation with a big budget, like an international TED Talk or an extravagant TV programme, then you can really go wild with this. There are so many modern possibilities, ranging from 3D holograms to flying into the theatre on a fleet of synchronised drones. Let's be honest though, most scientists don't have the budget of a Beyoncé concert to play with. So here, we are going to provide you with some realistic, accessible materials to jazz up your talk.

## 8.1    What Tools Do You Have?

Depending on the type of presentation you are doing, you may want to call upon support other than PowerPoint slides to help you out. So what else could you use? What other materials will you commonly come across? You may want to integrate films or other audio-visual clips into your presentation. You may want freedom to make live notes with a whiteboard, blackboard or a flip chart. You may even want to make your presentation really stand out by bringing in a sample or setting up a live demonstration. These methods can help you make your talk more interactive, stimulating and memorable.

## 8.2    Boards

In many locations where you will be presenting, there will likely be access to either a flipchart, blackboard or whiteboard. If there isn't, then you can often request one in advance. Being able to write things down as they happen enables you to pull ideas together from the room. For example, you can ask your audience's opinion on something and write a list of their responses. Active participation from your spectators like this keeps them engaged with your talk.

You can also add new bits of information to support you as you go through your presentation and come back to it at any moment. Be careful though, you should have plenty of ideas up your sleeve in case your audience is not motivated to get involved. Whether you have chosen to do a drawing or write some text on your board, make sure that it is sufficiently legible—without scribbles and in your best handwriting!

We divide the types of board you can use into two categories:

- erasable (blackboard/whiteboard),
- non-erasable (flipchart).

## 8.2.1  Erasable Boards

Whiteboards and blackboards are becoming rarer with the arrival of new technologies. You may come across a digital whiteboard, for example, but the concept is essentially the same. Just as with PowerPoint slides, you should not be using your board to write down paragraphs of text. Instead, this kind of tool is best used when describing complex ideas. When thinking about quantum physics or chemical structures, it can be extremely useful to be free to draw a schema or equation whilst explaining it.

Since the board is immediately erasable, you can very easily modify it or correct errors. But the disadvantage is that once you have filled the board, you will then need to erase it entirely before you can start anything else, meaning that you will not have a record that you can go back to later in the talk. This constraint is changing with digital whiteboards that now give you the option to save the image, which can be extremely handy. At the very least, you could always take a snap on your phone to come back to later.

In our training sessions, we use an example video of a volcanologist, Prof. Jacques-Marie Bardintzeff, explaining the different risks surrounding volcanic eruptions. He uses a whiteboard upon which he sketches a volcano and he adds content bit by bit. Whilst he describes the process of an eruption, he expands his drawing by adding keywords and imagery, building it up as the explanation goes on. It is a great way to help people follow his train of thought.

## 8.2.2  Non-erasable Boards

With a flipchart, you can come back to ideas later in your talk. Non-erasable boards are therefore more adapted to situations where you may want to recap or add to previous items, such as lists. Beware that exercises like this should not be totally improvised. Think in advance about what kind of ideas you expect from your audience. If your list is to help you make a point in your presentation, you should already have the ideas you need in your mind. On the other hand, if it is a true brainstorming session, you should avoid influencing their responses. Either way, be ready to be bounce off your audience and summarise on the spot.

Boards are the most interactive tools at your disposal because you can ask people to shout out answers or, when appropriate, get them to stand up and write things on their own. We saw a talk with a scientist speaking at an all-girls high school. At the beginning of her talk, she asked three teenagers from the audience to draw a scientist on the flipchart. Afterwards, she turned over the page and began talking about her own career as a chemical engineer. At the end of her talk, she came back to the three drawings. The girls from the audience had all drawn male scientists, some resembling the stereotypical, mad scientist with fuzzy hair and glasses, reminiscent of Einstein.

She then said, "And what about me? Have I convinced you that I am a scientist? I don't look like your drawings, but hopefully after today you will remember that women can be scientists too. That means any of you girls sitting in front of me today may follow a career into research. And if that is what you enjoy, I whole-heartedly wish it for you." It was a very inspirational talk and she managed to use her flipchart to make a hard-hitting point about women in STEM careers.

## 8.3    Objects

If a picture speaks a thousand words, then an object can speak a million. Not only does an object allow you to bring a visual aid out of a PowerPoint slide and into the real world for your audience to share, it also provides palpable evidence to support your message. We divide the types of object you can use into three categories:

- scientific,
- props,
- 3D metaphors.

### 8.3.1    Scientific

We worked with a researcher in the cosmetics industry who wanted to include a new active ingredient he had produced in a formulation. His project involved testing which kind of solvent would be needed for it to dissolve effectively. Like many chemistry experiments of this type, he simply added the same amount of his active ingredient powder to different solvents. Nothing too exciting. But to liven up his talk, he had hidden Erlenmeyer flasks with the various solvents behind a black screen. He only revealed them after he had described the experiment, to show how well the active ingredient dissolved in each solvent. He could have just shown some pictures on his slides but doing it this way was more theatrical and he managed to build up more suspense, making his results more impactful for the spectators.

## 8.3.2  Props

We consider props to be any object—scientific or otherwise—which is used to illustrate the (potential) outcome of a project or idea. A speaker in the defence industry gave a talk about the importance of developing lower-weight materials to be carried by soldiers in the field. He provided the example of how many kilograms a ground troop needs to carry daily. On average, the total comes to around 45 kg. He brought in the equipment that a solider is likely to wear in the field. He asked a member of the audience to try it on and made him walk across the auditorium. The person struggled a fair amount and the speaker explained that carrying more than 30 kg all day long can affect the cognitive, as well as physical, capacity of the soldier. In doing so, he successfully described a need to reduce the weight of each of the items—weapons, backpack and protective gear—by around 10–15 kg. The prop helped him emphasise how vital that is for soldiers who need to stay clear-headed.

## 8.3.3  3D Metaphors

This category includes any object that is not directly related to your work but helps understanding in the form of a physical metaphor. You can use everyday objects to make a point. We trained a speaker who was working on a docking system for space shuttle rendezvous with the International Space Station. He used a coin that was 2 cm wide to demonstrate the precision required when docking. In his introduction, he took it from his pocket where it had been hidden. He raised his hand so that everybody could see the round, shiny object. It was a clear, palpable example which supported his argument of how little room for error there is when it comes to space shuttle manoeuvres.

Last year, we worked on a scientific award ceremony where a series of scientists pitched their research projects to an audience of 300 peers. The principal investigators took to the stage to describe their topics with three minutes each to pitch. As a point of reference, each scientist brought an object with them as a figurative image. One researcher, who carries out epidemiological studies in ageing, used a watch on her wrist to represent time.

Another of our trainees worked on a mechanism used by bacteria to attack human cells. He had identified a biological system resembling a type of arrow that bacteria fire at target cells, to pierce their outer membrane and kill them. As a metaphor, he used a handheld, wooden bow with an arrow. It was a solid description, helping the audience to understand a complex molecular mechanism by comparing it to a more fathomable object. Not only did these objects serve to help researchers tell their story, they also helped the audience remember the research topics.

## 8.4  Demonstrations

As we described in Chap. 6, any talk will benefit from you showing something in action. There is nothing better when it comes to proving your point or convincing your audience that what you *say* works actually does. There is some risk involved here: if your demo does not work, it will have the opposite effect to that which is desired. A well-known car company learned this the hard way. They had designed a car with automatic safety breaks and decided to do a demonstration of the technology for an audience of journalists. In the demo, the driverless car simply sped towards the back of a stationary lorry to prove that the brakes would stop the car automatically as soon as it got too close. Only it didn't. The crowd of journalists ended up getting the scoop they weren't expecting. This left the car company pretty red-faced when journalists wrote about the safety system and how it *didn't* work.

Another of our trainees used a demonstration of a software development capable of detecting street violence in real-time using CCTV cameras. The researcher here had designed his new system with recognition software integrated into each camera to automatically detect violent movements. Upon sensing potentially forceful acts, a warning signal is sent to the authorities. During his talk, he set up a camera using his software that he aimed at the audience. On a screen, the spectators could see themselves and the way the image was being detected. Their movements could be sensed by the system and analysed as safe or not safe. Point proven.

Sometimes your demonstration may be too big or elaborate to bring into your presentation room. Examples we have seen include large robotic devices and wearable technologies, like VR glasses. Nevertheless, you can still have your demonstration either before, or ideally after, your presentation in the lobby or somewhere else nearby that is accessible and where people pass. That way, you get to do your demo and your talk will continue to live on well after you have left the stage.

# 8.5    Multimedia: Videos and Sound

## 8.5.1    Videos

If you want your presentation to be more dynamic, you can include video clips or short animations. You don't want your whole talk to be your audience watching a film, but it can help to make a good point. In addition to captivating the attention of your audience, video clips can help to effectively illustrate a point, or include a testimony from a fellow expert or even a kid who loves your new tech.

Do beware, clips can sometimes be difficult to find or costly to make. They are also prone to technical problems, so be sure to test them first in your presentation room if possible. It is wise to have a backup on your desktop or USB key. When showing a video, you also need to be aware of the sound, which can mask your own voice or make it more difficult for people to hear. To overcome this, either mute the video, set the volume very low and talk over it, or give your commentary before or after. If you have to compete with the sound of your video, it will confuse your audience.

A chemistry professor who was presenting to a live audience of young people, used video in an original way. He was describing how energy is absorbed by the intestine. To demonstrate this, he had actually gone to a gastroenterologist who carried out a capsule endoscopy on him. The procedure involves swallowing a small camera, in the shape of a pill, which makes a video of the journey through the digestive tract of a person. A few days later, in front of his audience, the speaker was able to show the film and say, "This is what is happening in my intestine."

Videos can also be used to help demonstrate a metaphor. We trained an engineer who was giving a talk about her research into additive manufacturing; a technique where objects are produced by adding layer upon layer of material, as opposed to cutting up different components and sticking them together in traditional production processes. Whilst her talk was to an audience of scientists at a conference, she wanted to add a twist of fun to her presentation. To help her explain how the process works, she showed a sped-up video of herself cooking a lasagne at home, adding the layers one by one. Later in her presentation, she showed a video of the process going on in her lab, this time producing a tank capable of resisting extremely high pressures. The first video provided her with a helpful analogy to better understand the second clip.

Sometimes you may have good video clips to hand. But that isn't always the case. Also, when using videos, keep in mind that you don't want them to be too long. Audiences tend to be very passive participants when watching videos and animations, so you risk losing them.

## 8.5.2   Animations

Speaking of animations, whether they are 2D or 3D, realistic or cartoons, they can be very useful for explaining something that in words or a single image would not be so easy. This makes them a great tool for explaining an idea which has yet to become a reality. One of our trainees was presenting his technological innovation to a crowd of fellow researchers and potential funders. His development was a kite, which could be used to haul boats out on the open sea, to reduce fuel consumption. He was lucky enough to have a 3D animation to hand showing how the system was expected to work. Since nobody had ever seen what he was describing before, the animation provided a comprehensible explanation of something that did not yet exist.

We trained a botanist who was studying the effect of climate change on plant life. He wanted to explain to an audience of journalists that plant species were moving in response to the change in their environment, due to global warming. He showed a small animation of the surface area of a zone in Europe where Beech trees grow. He had used his data to project where the trees were currently growing, compared to where he predicted they would be found in 2080. The animation showed the movement of the plants from one zone to another over this 60-year period, based on his results. It demonstrated a shockingly dramatic effect of climate change on the spread of Beech trees across the continent, helping him get his message across.

## 8.5.3   Sounds

Using auditory cues or examples can be a great way to integrate another sensory tool into your talk. There are many circumstances where a little sound bite can make all the difference. Your research may provide you with the possibility to integrate sound. We advise that if you can, do it! It is simple to use, is responsible for less technical problems than videos and can provide evidence to support your argument.

We worked with an urban planning team who had been designing housing right next door to a motorway. They had designed a barrier which they argued could reduce background noise by 10 decibels, quite a significant amount. The interesting thing about decibels is that they increase on a logarithmic scale, which means that the difference in terms of noise experienced is difficult to comprehend in your mind. A reduction in 10 decibels is actually more like a sound half as loud as the original—an impressive result. This concept is quite difficult for an audience to grasp. The speaker for the team blasted out the sound of traffic that would be heard from the housing block before and after installation of the soundproof wall. By doing so, the audience was able to better visualise (or 'audiolise') the difference, and so the presentation was much more convincing.

Other instances may be where you work on a technology and you want to demonstrate what it sounds like on the inside. Or maybe you work with animal calls, human communication, language, echolocation, heartbeats, the Doppler effect, the list goes on. We suggest you give it a try!

## 8.6 Before We Move on…

To summarise, we have talked about slides, flipcharts and erasable boards, video clips, samples and live demonstrations. This list isn't exhaustive, but it should give you some ideas about how to strike a chord with your audience—in an affordable way. Think about it in advance to ensure that the tools you use serve your message, be well prepared and everything will run smoothly. Look at the Table 8.1 for some of the key advantages and disadvantages of each technique.

**Table 8.1** A summary table of different supporting materials

| SUPPORTING MATERIAL | ADVANTAGES | DISADVANTAGES |
|---|---|---|
| POWERPOINT SLIDES | ATTRACTIVE, CLEAN | MUST ENSURE COMPATIBILITY |
| | PRACTICAL: CAN ADD PHOTOS, VIDEOS AND ANIMATIONS | |
| | PREPARED IN ADVANCE | DOESN'T LEAVE MUCH SCOPE FOR IMPROVISING |
| | SERVE AS REFERENCE POINT FOR AUDIENCE TO SITUATE THEMSELVES IN YOUR TALK | REQUIRES AN EFFORT TO BE CONCISE WITH KEYWORDS |
| | READILY AVALAIBLE, EASY-TO-USE | MUST VERIFY COMPATIBILITY WITH THE EQUIPMENT AVALAIBLE |
| BOARDS | HELPS TO GATHER IDEAS | NOT VERY PRACTICAL TO MOVE FROM ONE PLACE TO ANOTHER |
| | CAN BE BUILT ON PROGRESSIVELY DURING THE TALK | |
| | CAN COME BACK TO IT AT ANY MOMENT (IF NON-ERASABLE) | MUST WRITE NEATLY AND LIMIT TEXT TO KEYWORDS |
| | MAXIMAL INTERACTIVITY/ ADAPTABILITY WITH THE AUDIENCE | NOT MUCH CHOICE OF COLOR |
| | LEAVES ROOM FOR IMPROVISATION | REQUIRES AN ABILITY TO IMPROVISE |

(continued)

**Table 8.1** (continued)

| | | |
|---|---|---|
| **OBJECTS** | CONCRETE AND EFFECTIVE | SOMETIMES CAMERA IS NEEDED FOR A SMALL OBJECT, BIG AUDIENCE OR BOTH |
| | CAPTURES ATTENTION OF AUDIENCE AND AWAKENS CURIOSITY | |
| | APPEALS TO ALL FIVE SENSES | MORE APPROPRIATE FOR SMALL GROUPS, OTHERWISE AUDIENCE MAY LOSE ATTENTION |
| | CAN BE INTERACTIVE IF SAMPLES ARE PASSED INTO THE AUDIENCE | SOME SAMPLES CAN BE FRAGILE OR EXPENSIVE |
| **DEMONSTRATIONS** | LIVELY, CAPTIVATING | SOMETIMES EXPERIMENT SHOULD BE FILMED SO THAT EVERYONE CAN SEE |
| | BEST PROOF THAT SOMETHING WORKS | IT NEEDS TO WORK |
| | | NOT EASY TO FIND |
| **MULTIMEDIA** | CAPTURES ATTENTION OF AUDIENCE | CAN BE EXPENSIVE TO PRODUCE |
| | ILLUSTRATES A POINT OF VIEW | |
| | CAN TRANSPORT US TO ELSEWHERE OR BRING PEOPLE INTO THE ROOM | MUST NOT BE TOO LONG |
| | | MUST BE TIMED SO THAT THEY MATCH WITH YOUR SPEECH |

# 9

## Popularise Your Science

### Make Your Work Understandable to Everyone in Your Audience

When was the first moment you knew you were going to become a scientist? What was that click that motivated you into making a career in research? For many, it is the excitement of discovering new things about the universe that nobody knew about before. That sensation of moving humanity forward and contributing to the pool of knowledge. However it happened, you made it. You are now an expert on your topic, that is why you are the person presenting. Take a little moment to appreciate it.

As the expert, you know the science or technology you are talking about better than anyone. That has got to be a fantastic feeling. But being an expert isn't without its drawbacks; knowing more about your subject than anybody else in the world can sometimes be quite a lonely place. You know all these amazing things and can only share them with a restricted number of people. However, you can invite your audience to explore some of these exciting findings! Whether it be fellow experts or normal everyday people, your talk is the

© Springer Nature Switzerland AG 2020
A. Youknovsky and J. Bowers, *SELL YOUR RESEARCH*,
https://doi.org/10.1007/978-3-030-34181-7_9

opportunity to bring them into your world. Give them the chance to revere in your wonderful insights.

To do this, you will need to give them a helping hand. It took you years to know what you know now, and you have mere minutes to let them in. Before you can discuss your research with an audience, you must first explain it. To explain it, you need to fill in the gaps in their understanding.

Simplifying complex ideas is a tremendously difficult task. In a recent seminar about science communication, Laura Helmuth, Health, Science and Environment Editor at the Washington Post said, "It's hard to remember the words you used to not know. Try to capture that memory of being confused." One of the challenges of explaining your research to someone else is that you must be conscious of what they don't know. Here we provide you with tips on how to make your research more accessible and approachable to everyone.

## 9.1	What Is Science Popularisation?

Popularisation is about making your subject more understandable to different audiences. If the people you are speaking to do not understand the basics of your talk, they cannot and will not follow your argument. Consequently, you won't effectively transmit your message. In science, there is always some level of background knowledge required. It could be anything from understanding complex physics concepts like string theory, to seemingly more simple facts like knowing that apples have DNA in them.

Think hard about the needs of your audience in order to adapt your talk. For more help with this you can always go back to Chap. 3. Consider who they are and what facts or concepts they are coming to your presentation with.

### 9.1.1	When to Popularise

Bear in mind that simple does not mean wrong. As experts, you must be precise in your language, and most of the time that means being complex. The process of science popularisation can be a delicate balancing act. Too simple and your technical audience will feel frustrated and patronised; too complex and you may lose their attention. Here are some common examples of where your audience may not understand you.

---

**Example 1: Complex Jargon**

*"This technique generates over 100 kWh of energy. Better still, it is carbon neutral, which is great!"*

---

In Example 1, the terms 'carbon neutral' and '100 kWh' may not be explicit to everyone in the audience. What do they mean? How are they calculated? You should not assume prior knowledge of technical vocabulary such as this, because it could be misinterpreted. To overcome this filter, jargon needs to be defined.

For the quote given, your receivers would need to know that carbon neutral means offsetting carbon dioxide emissions, such as air travel, with an equal amount of carbon removal from the atmosphere, such as planting trees. Therefore, try knitting that information somewhere into your talk.

When using facts and figures such as 100 kWh, it helps to put a point of reference that makes the concept more tangible for your audience. It could be as precise as, "enough to power a 50-gallon electric water heater for around 10 days," or as vague as, "enough to power a large family home!".

---

**Example 2: Awareness of Current Events**

*"We are worried that this infection may spread as quickly as the recent porcine plague epidemic."*

---

The speaker in Example 2 is using porcine plague as a point of reference. They assume that their audience is aware of the recent epidemic of this disease, which may not be the case. What is it? What are the symptoms and risks? How dangerous is it? How many animals were affected?

The sender is also assuming that the receivers comprehend the gravity of the epidemic enough to use it as a point of reference to describe another event of a similar scale. Although the speaker uses the term 'quickly', if the receivers are unaware of this event, they will not be able to quantify the speed of this reference in their minds. Is it referring to a matter of days, weeks, months or years? If your audience are not specialists in your field, they might be stuck behind a filter of awareness of the current events you mention. To improve the talk, the speaker in this case can do one of two things:

**Change the point of reference:**
   *"...spread as quickly as the Influenza virus during winter months..."*

OR
**Give more details to quantify the lacking information:**
*"...as quickly as the recent porcine plague epidemic, which moved from Russia to Belgium in just over a decade..."*

The former overcomes the filter by using a more common example, whereas the latter provides additional information to better illustrate the case described. Either way, by modifying the points of reference used, the speaker is helping the receivers to get the message more easily.

Wording that you use can have different meanings to different people. Almost all scientists would accept the word 'significant' in Example 3 to indicate that the results obtained in a study were verified and validated using widely accepted mathematical models. In short, to scientists the term 'significant' means 'statistically significant'.

---

**Example 3: Terms that Mean Different Things to Different People**

*"We measured the properties of an optical implant, capable of running extremely complex mathematical calculations. Our results were **significant**."*

---

To a non-scientist however, 'significant' has a different connotation. When used in everyday language, it refers to something that is important or meaningful. Thus, if you use that term in a talk with non-scientists, then the quote above takes on new meaning: the results were important.

That not only changes the meaning of what is being said, but it also begs the question: why is it important? And how is it important? Perhaps specifying that the results were verified by statistical tests, proving them to be valid, would be a better fit for a general public filter. Otherwise, you can leave maths out entirely. Most non-scientists are not familiar with probability statistics and so more often than not there is little added value in even mentioning them.

## 9.1.2   Are You Making Yourself Clear?

During your presentation, you can measure the attention of your audience by looking into the eyes of the people in front of you. If they are watching you and look concentrated, it means you're doing great. If most of them are playing with their phone or staring off into the distance, they might be lost. It is never too late to react if you think your audience has disconnected. Try asking a question to check if they have understood you. It will help wake them up. They may need you to go back over your explanation, so give it another shot if necessary.

### 9.1.3  Signpost Your Talk

Through your talk, take your audience by the hand. Keep reminding them where you are in your presentation, particularly in circumstances where content is difficult to understand. To help, you can use transition words. Terms such as: "first, second, third", "if we combine all four factors I just mentioned…" or "as a consequence", can really help to guide your spectators. You can let them know that you are changing topic or if you are continuing down a list. Of course, before the end of your talk, a classic "as a conclusion" or "to conclude" will be welcome. Following your trail of thought will help them stay engaged.

### 9.1.4  How to Popularise

Over the years we have built up a collection of rules to follow to do a good job of popularising your research. Our guidelines fit into four categories, which we refer to as the three Cs:

- make a connection,
- give context,
- be concrete.

You should take a look through them and, depending on the specific nature of your presentation, reflect on which ones will help you get your point across in the most effective way.

## 9.2  Make a Connection

Proximity with your audience will keep them engaged for longer. As a speaker, you need to create a bond with them. This can take several forms: technical, personal and humorous.

### 9.2.1   Technical Impact

On the technical side, your spectators will need to know how your work is relevant to them. For example, at a presentation about personal data in front of a general public audience, a speaker started with, "Who here leaves location services activated on your apps even when you aren't using them?". He gave them a little time to raise their hand and asked, "Are you sure?", with another pause. The silence here was intriguing, because his point was that most people don't know for sure. He then listed the most famous apps which collect location data that we aren't even aware of most of the time. With two simple and well-delivered questions, he emphasised the relevance of his topic to his audience.

### 9.2.2   Personal Insight

On the personal side, scientists often find this a bit harder. In the world of research, people frequently shy away from giving up personal details about themselves, letting the data speak for itself. This is actually a huge mistake. Your audience cannot listen to data. They must listen to a human being: you.

Whoever you are, you have feelings and emotions, qualities and flaws, just like the rest of us. Don't be afraid to let your audience see that. Give something away about yourself. Why are you so passionate about your subject? How many times did that experiment fail before it finally worked? These things are personal but not too intrusive. Just enough for the people in front of you to feel a connection with you. If your audience is on your side, they are much more likely to listen to what you have to say. You could talk about personal experiences you have had, trips you've been on, people you've met and so on. Usually, American speakers naturally talk more easily about their partner or their childhood, whereas Europeans find it harder to be that personal. Do what you feel comfortable doing.

### 9.2.3   Funny Tales

Science talks can so often be very serious and dry. Everyone loves to laugh, and everyone can make a joke. The most important thing about humour is that you use gags that you are comfortable with, otherwise it can very easily flop. One of the safest types of humour is self-deprecation, as it is based on you laughing at yourself rather than something or someone else. But any type of humour, as long as it is not offensive and matches your personality, is worth a try. If your audience doesn't laugh when you thought they would, simply move on and let the joke go. Over time, you will learn what works for you and what doesn't. But if it functions and your spectators do laugh, leave a silence for them to enjoy the moment. Don't step in on your audience's laughter.

## 9.3    Give Context

As we said in your mind map plan, your audience want to know the context of your work. Why are you even bothering to research such a topic? This means highlighting not only the scientific challenges you are working on, but also the social, economic or even cultural relevance. When you tell your audience the reasons for doing your research, you also need to tell them why they should care. Explain how your research can be useful to them and how it will affect their life. Don't assume that your audience loves your topic as much as you do.

### 9.3.1    Choose a Character

You are essentially telling a story about your work and any good story has a challenge that the main character must overcome. Tell us why! We recently trained a group of scientists in public speaking for a Trust who had funded their research. The topics ranged from economics to biology to geology. We challenged each speaker to summarise their research in a pitch no longer than two minutes. The only other constraint was that they should include a person in their talk—real or fictional—who will benefit from their research in some way. All research has an intended outcome, and that is almost always related to a human need. If it's biology, it could be a medical treatment. Seismology, predicting when tsunamis may occur to protect communities living nearby. Economics, helping farmers in developing countries to access banking services. Who is your main character?

### 9.3.2    Make It Relatable

Let's take an example from the field of medicine. A researcher has designed a biosensor for cardiac monitoring. His innovative wireless device allows anybody wearing one to track data relating to their heart function. Whilst this is all well and good, to make the topic relatable to everyone and not just people worried about heart problems, the researcher could inform the audience of how many patients' lives will be improved because of this technological development. Or even estimate the savings that will be made by health services as a result.

For an event we worked on, we asked the speakers to focus their talks on the three of the current Sustainable Goals defined by the United Nations. They served as a link to pull the varying projects that were presented together. The group of around fifteen researchers explained how they aimed to contribute to achieving these objectives, which include taking care of the oceans, tackling gender inequality and providing access to clean water for everyone. To do this, we asked them to tell their audience about the world they hoped to be living in by the year 2030. In marrying the overall context of their studies with a justification of their commitment, each of the talks became much more personal and therefore more touching for the audience.

## 9.4   Be Concrete

Janna Levin, a physics professor and science communicator said, "Ambiguity is very interesting in writing; it's not very interesting in science." Whilst she was referring to the difference between a literary novel and a piece of science writing, the same applies to oral communication. The goal of your presentation is to successfully transmit your message to your audience. Rather than have them walk out of the room pondering on the meaning of your talk for the next few days, you want them to leave knowing exactly what the importance of your topic was. You should be clear, unambiguous and explicit about the information you are giving.

### 9.4.1   Jargon

Regarding technical terms, try to limit the use of jargon to only those terms which are necessary to make your message clear. The essential technical terms will need to be clearly defined. The rest: forget them! Use everyday words and phrases. Make it as easy as possible for anyone to understand you. We repeat, simple does not mean wrong.

If your audience may be unfamiliar with an acronym, you need to define it. If you can give your talk without it, cut it. Banning acronyms and abbreviations will make your presentation more intelligible. Even if it's written in a legend on your slide, people who are unfamiliar with the acronym will need time to compute what it means. Also, the same acronym can often be used for different purposes depending on the field. PCR in biology means polymerase chain reaction, a common molecular analysis technique. Whereas in psychology it refers to Parent-Child Relationship, a common interaction studied by psychologists. In business, PCR is a Project Cost Report.

A metallurgy expert was presenting his work on the development of a "3D multi-physical model to assess the weld mixing of advanced high-strength steels for use in the automotive industry." This particular type of alloy reduces car weight and increases mechanical performance. For a talk to a lay audience, he needed them to know that steel was an alloy, that it is used in the automotive industry and that he was working on a high-strength, lightweight version of the metal. Neither the term 'advanced high-strength steel' nor its subsequent acronym AHHS, nor the expression '3D multi-physical model' were necessary for a non-specialist audience. However, for an expert audience, he would have needed to go much further the other way. He would not only need to mention AHHS but also that his work was related to the computer modelling of 3rd generation AHHS. For experts, this jargon is necessary to go into the finer details of his research.

## 9.4.2  Data

Be tangible. Use data which refer to universal notions and that are self-explanatory. It is extremely difficult for most people to visualise how quick 1,798,754,748 m/s is. Many wouldn't even know if it's incredibly fast or not, so the point you were trying to make would go straight over their heads. Instead, you could say, "Six times the speed of light." Most people know that the speed of light is fast, so six times this must be really fast.

Graphics need to be simple. A bar graph, for example, should not include more than five or six data points. If your graph has four bars but you only refer to two of them, cut the extra two bars out. Whilst for an expert audience you may be able to take graphs directly from your paper, you should consider transforming them completely if you are giving a talk to a non-specialist or general public audience. That means producing new, more basic graph types. For example, pie charts are rarely used in research papers, but they are great for visually describing data to a general public audience.

Also, don't be afraid to drive your point home. For statistics that are particularly important to your message, you can often use two turns of phrase to describe the same data. For example:

*More than 9 in 10 Australian adults did not eat the recommended amount of vegetables in the year 2014 to 2015. That means less than 10% of over 18 s across the country are eating enough vegetables.*

For the most important data relevant to your message, you can use repetition like this to drill numbers into the audience.

### 9.4.3    Visuals

Complex ideas can often be more easily understood in the form of an image or a diagram. But be careful that they themselves aren't too complex. If you must explain it, then it's probably too complicated. To provide you with photographs and video content, you could try taking snapshots whilst working. Even photos taken on your phone can be used in a presentation. You should ensure that they are of good quality in terms of definition and lighting, you never know when you could use them.

## 9.5    Try It Out…

Our three Cs should set you off on the right direction for popularising your talk: connection, context and concrete. If possible, run through your presentation with people who have the same level of knowledge as your audience will have. Their feedback will help you gauge whether you are popularising to the right level. Your audience's reaction to your talk will also help you to identify ideas that you could reuse or rework for your next presentation.

# Part III

## Third Step: Breathe Life and Fine Tune Logistics

# 10

# Master Your Non-verbal Communication
## Use Your Body to Convey Your Message

Look at the photograph in Fig. 10.1. Can you guess who is lying?

The answer is: no. Why? Because we do not know what the people are saying. Whilst we can make a snap judgement about their feelings based on their body language or facial expressions in the picture, we cannot possibly be sure what that says about their truthfulness. We can only estimate their state of mind based on what we see. Are they comfortable? What emotions are they feeling? Are they engaged with one another? As humans, we have the ability to recognise these things when looking at other people. It is an important skill required in social animal species such as ours.

Therefore, to fully understand a situation we need more information. This is because in communication the non-verbal and verbal components combine to provide us with a complete picture. Imagine if in the moments before this picture was taken, the woman in the pink shirt had announced that her daughter had been taken to hospital that morning. Does she look like she

© Springer Nature Switzerland AG 2020
A. Youknovsky and J. Bowers, *SELL YOUR RESEARCH*,
https://doi.org/10.1007/978-3-030-34181-7_10

**Fig. 10.1** Three people in a meeting. Who is lying?

is daydreaming about that? On the other hand, what if she had just tried to speak up and the two men had ignored her? Are you still reading her body language the same way? Some experts may be able to pick out more information than us everyday folk, but everybody can detect these types of signals, even if only subconsciously.

Our body speaks for itself. It has a whole language of its own, most of the time without our knowledge. Non-verbal communication includes eye contact, gestures, tone of voice and breathing. Since these responses are mainly reflexes, you cannot completely master them. Hence, one of the biggest questions in public speaking is how to manage these signals rather than overcome them. Too much control will be visible to an audience and will inhibit your natural demeanour.

The way to master your non-verbal communication is to put yourself in the best possible condition so that you can fully commit to your content, express your excitement for your results and convey the right emotions at the right time. You built a presentation in the first two steps of the SELL Method, but so far, your preparation has been purely intellectual. Now, let's now get physical! In the third step of our SELL Method, where the first L refers to life, we will concentrate on how to deliver your presentation and exercises to improve yourself.

# 10.1 What is Non-verbal Communication?

Non-verbal communication refers, as its name suggests, to everything that we say without using words. It relies on two primary media: presence (everything an audience can see) and voice (what they hear). In the world of communication, we talk about congruence, which refers to the harmony between what is said using spoken language and the way it is implied using non-verbal communication. For the audience, congruence is achieved when all the signals they receive tell the same story and so reinforce one another. This makes us trust people and connect more easily with what they are saying. On the other hand, a lack of congruence weakens a speech and incites suspicion.

Non-verbal communication includes physical elements (posture, attitude, gaze and facial expressions), auditory signals (like tone, volume, rhythm and speed of speech), or equally important subtle effects such as breathing and silence. This multitude of information combined with spoken words influences the way your spectators feel.

Let's say you are sitting in the audience at a talk and the presenter starts smiling whilst telling a sad story. Instinctively, you realise that the person lacks congruence between his verbal and non-verbal communication, so your mind begins to seek meaning in that disconnect. Perhaps this detachment prevents you from trusting the speaker, or maybe you feel that he is being ironic and therefore take his speech as comical. However, he may also simply be nervous standing in front of a group of people, and his smile is his underlying discomfort showing through. From the outside, it is difficult to determine the cause of the disconnect but the consequences of non-verbal signals are nonetheless visible and affect your judgement of the speaker.

Unfortunately, it is impossible to completely control our body language. Our body doesn't ask our permission to do what it considers best in a given situation. There are, however, techniques that you can use to improve your non-verbal communication. Hence, the best advice we can give is the following: to inspire confidence in others, you must first have confidence in yourself, your project and how you present it. For that, you should feel that both you and your presentation are good and ready! Before a deadline, we advise you to practise your speech several times out loud and use some of the following exercises to polish your presence.

## 10.2   Voice

Your voice is undoubtedly an important tool in public speaking. The way you use it provides a bunch of information to the listener. Every person has a natural pitch and tone. However, in stressful conditions it may become higher or you could easily find yourself talking very quickly. Below we describe different aspects of your verbal communication that you can work on during rehearsals to elevate your delivery.

### 10.2.1   Breathing

Breathe: this is a highly important point. When you are nervous, you may find yourself breathing in at strange moments during your speech. Your audience will be aware of your flow of air in and out of your lungs, as humans have a habit of synchronising their breathing subconsciously with others. Hence, your erratic breathing can make the people in front of you feel uncomfortable.

Thankfully, there are some very easy things you can do to take care of this. Firstly, take some deep breaths before you begin. It's worthwhile doing this before you are called to the stage. Many people's anxiety is at its highest when they are anticipating their name being called out. Once you are on stage, breathe before you start talking. This will help set the rhythm and speed for your talk.

### 10.2.2   Speed

How fast you talk affects both the audience's enjoyment and understanding of your presentation. So often we have been to talks by scientists who took to the stage and rushed through their speech at lightning speed. Subconsciously, people try to escape the situation—if they speak faster, it will be over quicker. Nerves are normal, but if your audience can't understand your content, you may as well have never stood up in the first place.

Inversely, whilst it's generally less of a problem, some speakers tend to talk very slowly. This can have an energy-sapping effect on the audience. At least a bit of speed bring with it some liveliness. When rehearsing, practise getting the momentum right. Try using exercise 1 to set the pace. It will be set once you get through the first few slides of your presentation, so focus on your speed early on during your introduction, and then you can relax.

**Exercise 1: Slowing the pace**

Do you talk too quickly when you present? Practise the first two minutes of your talk out loud very slowly, adding pauses every time there is a punctuation mark—a full-stop, comma or brackets. Forcing yourself to slow down in this way will make you more aware of your speed. Then try again without this conscious pausing, and you will find that your pace is much more relaxed.

### 10.2.3  Tone

You need to deliver your words with a sense of conviction. If you are telling us that you are angry about the amount of plastic in the oceans, we need to sense it from the tone of your voice. If you are happy about discovering a new result that might contribute to the development of vaccines against HIV, then act like it.

Either way, the best way to be congruent between your voice and your words is to genuinely believe what you say. Real conviction is impossible to replace. If you feel concerned or passionate, your audience should feel it. Don't be afraid to be more demonstrative that you would normally be. In front of a large audience, it will seem less over the top than you think.

### 10.2.4  Volume

How loud you speak will have a direct impact on comprehension. In a large room, if you speak too quietly then the people at the back will not hear you. After all this preparation, that would certainly be a shame. If in doubt, ask the audience. Do be sure, however, that if they say no, you increase your volume—otherwise you could end up with a frustrated back row.

On the contrary, you may be in a small room or an area that echoes. In these cases, the opposite is true. If you speak too loudly, your talk will be uncomfortable to listen to. If in doubt, a little too loud is better than a little too quiet. In many cases you will probably get away without the need to adjust your own volume because you will likely be given a microphone. Obviously, make sure you keep it near your mouth and don't shout. For more on microphones, see Chap. 12: *Perfect Your Performance.*

### 10.2.5 Diction

Articulating correctly can make a world of difference, especially in science where your audience is often international and so English is likely to be their second or third language, as for yourself. It is disconcerting when a speaker doesn't pronounce a word very clearly, and so spectators are unsure of what it was. Whether you are a native English-speaker or not, you should pay attention to your diction (Exercises 2 and 3).

---

**Exercise 2: Voice training**

Another way for improving your delivery is to combine the previous exercise with one using music. This task is regularly used by actors who want to familiarise themselves with their lines and find new ways of saying them. It will help you to improve your articulation and find a rhythm that is not monotonous.
1. First, practise your presentation out loud whilst over-articulating every word.
2. Next, sing your presentation—this is easier to do with a melody you make up on the spot rather than a song you already know.
3. Read it out normally.
   You should find that on the third time, you will be speaking clearly without too much effort. Give it a go! Of course, you can do it on your own to avoid embarrassment if you wish.

---

**Exercise 3: Articulate**

To improve your diction, you can try a sequence of over-articulating your text. Accentuate the movements that your mouth makes around each syllable. Try to stretch your muscles as hard as possible. If you have a problem with articulation and people have mentioned this to you in the past, we suggest you rehearse your presentation with a pencil in your mouth, placed horizontally between your teeth. Make sure you can still be understood. Then, when you practise again without the pencil, you will find that your diction is much clearer. An old classic, but it works well!

---

## 10.3 Body Language

Posture, gestures and facial expressions are all components of body language. They can indicate whether you are happy or sad, comfortable or anxious, or even lying. You can also use them to better express your ideas. Moving at just the right time or expressing emotion with your hand gestures can have a real impact on the audience's experience.

## 10.3.1  Posture

Studies have shown that your posture is not simply a 'symptom' of how you feel: you can have an open posture because you feel confident, but you can also feel confident because you have an open posture. The way that you hold yourself not only affects your audience; it also plays a role in how you feel about yourself.

In an intimidating situation like giving a presentation, speakers sometimes want to curl up and hide, resulting in them slouching, hunching their backs or leaning on anything around them: backs of chairs, tables, pulpits etc. To improve posture, keep your hands out of your pockets and avoid folding your arms or putting them behind your back. Instead, position yourself in a more solid stance. Stand up tall and straight, raise your head up high with an open body—this has been proven to make you look and feel confident. It also helps you get more of that precious air into your lungs as you stick out your chest, opening your ribcage as in Exercise 4.

---

**Exercise 4: Imaginary rope**

When you practise your talk, imagine a piece of string attached to the top of your head, pulling you up towards the ceiling. You should feel your neck stretching, your shoulders opening and your chest pushing out. This will improve your overall position and give you a more confident posture, that will be as effective on the inside as it is on the outside.

---

## 10.3.2  Gestures

Your body, particularly your hands, is a great tool for helping you to add depth to your words. You can use your fingers to point in a certain direction, describe the shape of a curve or hold up numbers. Whilst we are not suggesting that you mime your entire talk, it is worthwhile using your hands to help you tell your story. Your gestures can also help you to project emotion. If you are describing something that makes you angry, let your body express it; maybe your hands will move in a very solid way. If you are describing something light and airy, then your gestures should be smooth and relaxed. These actions will influence the atmosphere of the room.

When you are rehearsing, pay attention to what your body does and how it moves. Even the best of us have an annoying gesture that we find ourselves doing when we are nervous. Some people click their fingers, others twirl their hair or play nervously with a pen. Find out what your tics are and focus on ironing them out of your performance. Don't hold anything in your hands that you don't need for your presentation. You may have a microphone or presentation remote but slips of paper or fluff from your pocket are unnecessary. Concentrate on using your body to represent what you are talking about. Try Exercise 5 to help you connect your mind and body.

---

**Exercise 5: Physical preparation**

Stretch your neck, shoulders, arms, legs and joints. To work the muscles in your face, pull a grimace several times and then relax. This will prepare your body for action and will help your mind to stay connected to your physical form, which is vital for maintaining good presence.

---

You can use a technique that we call 'placing objects in the air'. If you are describing several things at once, perhaps how they interact, you can use your hands to drop imaginary shapes into the space in front of you. For example if you want to discuss a journey from point A to point B, you can point to a spot in the air in front of you for A, then another for B. If you do this one after another, your audience will create a mental image of their position and remember where you 'placed' them. You could then draw an imaginary line with your finger between the two. Oddly, you have created a sketch in your spectators' minds with nothing but movement and empty space.

## 10.3.3   Facial Expressions

Have you ever read the frustration in someone's face from nothing more than a minute, unintentional eye-roll? Or a light tightening of their lips? We can read a massive amount of information from facial expressions, even tiny—almost invisible—muscle movements, known as micro-expressions. In the interest of congruence, your face should be in sync with what you are saying. But don't overdo it: you run the risk of looking ridiculous. Again, the best thing to do is to connect with your own emotions. Your facial expressions will follow in their own way.

### 10.3.4 Movements

People often think that they must walk around the stage during a presentation, but movement is actually not a necessity. More importantly, you should make sure that you are stood with your feet firmly on the ground (with your great posture we just talked about) facing towards your audience. This is your default position. You can stay here for the entirety of your talk if you don't feel comfortable moving, relying on your body language to do the work.

If you do decide to add movements, they should add meaning to what you are saying and be thought out and direct—not tottering around the stage. Some people tend to rock from foot to foot when they are nervous or trying to remember their lines. Even though they themselves may not be aware of these movements, it is very uncomfortable for the audience to watch.

Also, remember that your face should always be in view of the audience. Don't turn to look at a PowerPoint slide. Not only will your voice become more difficult to hear, but you will also lose your connection with your spectators. If you need to see your PowerPoint slides, place your laptop in front of you so that you have no reason to turn around.

## 10.4 Eye Contact

Research into ethology has shown that in the animal kingdom, including primates, the 'stare'—looking at another animal with an unbroken gaze—is reserved for two types of situations. Between two members of different species, it is used by predators to concentrate on their prey. Between two members of the same species, it is a sign of aggression or a challenge. Generally this insistent gaze is brief, as one of the protagonists will lower its eyes. It's therefore not hard to understand why a whole room of people staring straight at you can be stressful.

When presenting, eye contact is important if you want people to listen and engage with you. By looking at the people in your audience, you are connecting with them. It feels unnatural to everyone at first. It's intimidating. Practice can help though. Rehearsing with other people would be great—if not, try imagining lots of people around the room and looking at them in their imaginary eyes.

Chris Anderson, curator of TED, suggests that you find a few friendly faces in the audience: those are the people you will look at when talking (see Exercise 6). It's a nice technique as it will help you feel calmer too. Also, try to look at people all over the room, front and back row, right, centre and left, not just in one corner. Use eye contact to spread your gaze across the whole area.

---

**Exercise 6: Improve eye contact and silences**

In Exercise 1, we presented you with a technique to help you slow your pace using punctuation marks. Here, we propose a variation which will also help you to improve eye contact with your audience. To do so, take any text you have to hand—if your presentation is written out then you can try with that. You should read it aloud three times:

1. Read your text whilst saying each punctuation sign in a loud voice ("comma", "full stop", "question mark" etc.), imagining that you are looking at someone different every time.
2. Read your text again and, on each punctuation sign, imagine that you are looking at someone different, say his or her first name in a loud voice.
3. Finally, on each punctuation sign, imagine you are looking at someone different, pause for two seconds, silently.

This exercise effectively combines silence and eye contact, adding strength to your words.

---

Whilst this may seem like a lot of different things to do, we suggest that you try them all and pick the best combination of exercises that works for you. The goal isn't to turn you into some sort of non-verbal guru, but to find a way for you to feel calm enough to be energetic about your own content because that is all that will ever be expected of you. If your personality shines through, then your audience will be on your side!

## 10.5  Before We Move on…

In a good presentation, the speaker is living in the present moment, not thinking about what just happened or what is about to happen. That means they can exchange and improvise on the spot if needed. They know their talk so well that they are not thinking about it anymore, just delivering.

This is what sportspeople and artists refer to as 'flow'. When you are in your flow, it feels as if time no longer exists, everything is happening on its own, and your body is doing the work for you. That feeling is not only amazing, but it is the most sure-fire way to have an impact on an audience. The more prepared you are, the closer you will be to achieving your flow.

# 11

## Manage Your Anxiety
### Calm Your Nerves on the Day

A young starlet once announced to the very famous French actress, Sarah Bernhardt, "I never get nervous on stage". Sarah replied, "Don't worry, my darling, it will come with talent." Nerves are normal when standing up in front of an audience. In fact, more than that—they are *essential*.

The time has come for you to take the stage. Your heart starts to pump blood around your body. You feel a little light-headed when your name is called by the chairman and you may even have the impression that you have forgotten everything you were going to say. Don't worry. We all get nervous when standing up in front a group of people, especially if they are likely to criticise our work.

As empathetic humans, we root for people who openly expose their weaknesses or who aren't afraid to laugh at themselves. Don't try to be something that you aren't, your audience will recognise the falseness straight away. It

© Springer Nature Switzerland AG 2020
A. Youknovsky and J. Bowers, *SELL YOUR RESEARCH*,
https://doi.org/10.1007/978-3-030-34181-7_11

sounds like a cliché, but just be yourself. Accept your flaws and your audience will too. At the end of the day, these are human beings sitting in front of you, not velociraptors!

Let's be clear: you cannot avoid the rush of adrenaline that you will get from standing up in front of a bunch of people, many of whom you may not know. But this adrenaline is exactly what will help you get through the experience. Still, there are simple things you can do to make anxiety less of an issue, without removing it entirely. This chapter is here to help you reduce that stress.

## 11.1  Practise, Practise, Practise

The most fool-proof way of reducing your anxiety to a minimum is practice. You should go through your presentation, learn what you are going to say and when, and rehearse. Practise in conditions as close as possible to those you will have on the day. Stand up at the front of a room with your slides (if you have some), looking out over a real or an imaginary audience.

### 11.1.1  To Learn or Not to Learn

As we have already mentioned, there are several ways to prepare your talk in terms of your text. Some people prefer to have text written out word for word in advance. We would particularly advise this for a short talk, such as a pitch competition like FameLab, where time constraints are at their tightest. Others prefer to work with an outline and then fill in the words as they feel it. Either way, you need to practise your text out loud and time yourself. Use a stopwatch so that you are aware of how long you expect to be.

Yes, you are no doubt busy with a million other things, but you can *always* find time to practise. Carry a copy of your notes around with you or put a printout of your presentation in your bag so that you can flick through them on the train home after work, at the bus stop, in the shower or in the waiting room at the doctor's surgery. Practise saying your lines to yourself, especially the opening paragraph—if you start well, the rest becomes much easier (see Exercise 7).

**Exercise 7: "Une Italienne"**

Theatre actors in France use a technique called *une Italienne [an Italian]* when practising their lines. Generally, before they go on stage, they will run through their lines out loud with a neutral voice and tone, helping them memorise their text without tiring themselves out. We suggest you do the same when rehearsing. If you have your text written in advance, then practice saying the lines from beginning to end. If you haven't written down your text, this exercise will help you work on your speech in advance and find the words you want to use, even if you don't learn them by heart.

Actors swear by this method because once you know your lines inside out and back to front, that's when you can really play with them. It also means that if for any reason you get cut off, you can jump back in at any point with ease. Distractions will no longer be a threat to your confidence! This can also help you articulate the words correctly, which is a must-have if you are not a native speaker of the language.

You will surely know people willing to help you rehearse. There are different people you could ask, depending on the kind of feedback you need. Get your colleagues who know your subject if you want advice about your content. Ask a friend or family member if you're looking for comments on your body language, delivery and level of popularisation. Or why not do Exercise 8 and film yourself.

It is time consuming, but the investment is worth the energy—especially for important, high-impact presentations. Remember: the harder you work, the less it will show. And the more likely you are to reach your 'flow'.

**Exercise 8: Film Yourself**

Film yourself giving your presentation and watch it back. Most people cringe at the thought, but it's a really good way of seeing how you appear to others and giving yourself some constructive criticism. If you have someone who can help, get their opinion too. You can film your talk on your phone and, if you're short on time, watch it on your way to work or your lunchbreak.

Filming yourself is a very useful technique, but beware that we have come across two kinds of response when using it in our training sessions. Some people find themselves better than they expected, and so are reassured. Others judge themselves more harshly than their audience would, because they are overly critical of the way they look or sound. Nevertheless, most people feel more confident afterwards.

## 11.2   Enter the Stage

You may think that the first moments of your talk are when you start speaking. In fact, your presentation really starts just before that, as you enter the stage. Depending on the context of your talk, this could be an actual stage, the floor of a lecture theatre, the front of a meeting room or any other setting. In some cases, you will be introduced by a chairman, host or previous speaker. In others, you will be responsible for taking to the stage and beginning your talk of your own accord. Either way, how you enter your stage will affect the minutes that follow during your presentation.

Before taking to the stage, we suggest our trainees use positive visualisation to help calm their nerves. Gather a pleasant memory in your mind: a calm, sunny beach from a relaxing holiday, for example. Close your eyes and hold onto that memory. It will help you smile and feel positive.

When walking to the podium, don't rush. Once you're on stage, take a pause and position yourself facing the audience. Then look at them, welcome them with a gesture and smile. Don't speak until you're ready, waiting for two or three seconds more than you are used to. This pause will give yourself time to adapt to the room before you begin. It will also signal to your audience that you are ready, and they should listen up. If you start calmly, then you are much more likely to stay that way for the rest of your talk.

## 11.3   Breathe

What happens when you are nervous? Your head may become fuzzy, your heart beats faster or you may sweat, to name a few. Unbeknownst to you, your breathing has changed. To get your airflow back to a good rhythm, feeding your brain precious oxygen, take a few minutes to do some abdominal breathing. This means breathing into your belly, as opposed to just your chest. It is used by singers, actors and musicians to help them get more air than usual into their lungs. You can do this anywhere, even in your seat in the minutes before your talk (see Exercise 9).

---

**Exercise 9: Abdominal Breathing**

For this breathing exercise, we advise you to find a place that is calm and isolated. Depending on the space around you, you can do it standing, sitting or lying down. Although not obligatory, we advise you close your eyes for a deeper relaxing sensation.

---

First off, you need to find your abdominal breathing. Concentrate on the air around you. Place your hands on your tummy, then breathe air in through your nose. Force the air into your abdomen. You should feel your belly expand. Once you've filled it up, breathe heavily through your mouth and feel your tummy shrink.

Once you are in the rhythm of breathing in and out of your abdomen, you can then start the following cycle:

- Breathe in through the nose for a count of four,
- Hold your breath for a count of four,
- Breathe out of your mouth and empty your lungs of all the air for a count of eight. The flow of air may make noise whilst you're doing so.

You can repeat this cycle three to six times. Afterwards you should feel more relaxed and capable of smiling.

You should also focus on your breathing in the first moments of your talk, in order to start well. We have seen talks where people were desperately fighting to catch their breath as if they had just run a 400 m sprint. If at any point you find yourself a little breathless, don't be afraid to stop for a few seconds. Take a deep inhalation and concentrate on speaking slower.

# 11.4   Project Yourself

Successful speakers dedicate enough headspace to thinking about their talks in advance, and they prepare themselves for every eventuality. When you limit last-minute problems, your anxiety levels stay where they should be.

When we are nervous about something, we try not to think about it too much because it frightens us. Yet, we advise you to visualise the day of your presentation well in advance. Imagine yourself arriving at the event, taking to the stage, giving your talk and so on.

By doing so, you will be able to ask yourself all the questions that need to be answered before the day of your talk. This will help you to be as comfortable and stress-free as possible. Your presentation will never be as good as it could be if you turn up ten minutes late, drenched from the rain having left your USB key in the car parked down the road.

Here are some of the questions you can ask yourself as part of this exercise:

- What time should I wake up?
- What will I wear?
- Can I iron my shirt the morning of my talk?
- Should I take my outfit with me and change on location?
- How can I get to the location?

- How long does it take?
- Will I need some cash?
- Do I need to bring ID?
- Will there be someone to meet me?
- Can I bring equipment?
- Is there somewhere to eat on location?

These questions are different from the questions we set out in Chap. 3 because they require you to visualise how the day of your talk will go. Knowing these things will lower your anxiety.

## 11.5   Respect Your Body

Healthy body, healthy mind. In order to be at your peak performance when put under pressure, you should take good care of your body. After all, it will be doing a lot of the work! When you find your 'flow', your body may switch to autopilot, so best make sure it is in good shape. When we are anxious, we are completely locked away in our heads and this can lead us to forget about our physical needs.

There are things you could do to keep your body at its best for your presentation:

- The adrenaline rush from presenting will make your body use up lots of energy, so make sure that you have eaten enough on the morning of your talk or around an hour before. Don't eat right before if you can avoid it, as this can lead to drowsiness.
- Following on from the previous point, try not to turn to high energy snacks or drinks just before your talk. The energy they provide is used up extremely quickly by the body. You don't want to have a sugar crash at the halfway point.
- Tiredness influences the efficiency of our brain, negatively affecting memory. It also reduces our ability to think fast on the spot, a skill needed in order to improvise. Try to get as much sleep as possible the night before your talk. Go to bed early (for once).
- Another point regarding tiredness: alcohol intake. If you are at a three-day conference, you may be tempted to have a few beers or glasses of wine with dinner the night before your talk. Try to limit your consumption.

- Caffeine can induce anxiety as it speeds up heart rate. If this happens to you, we recommend that you avoid coffee just before your talk, even if you need an energy boost.
- Try to get some exercise in the day before or the morning of your presentation or listen to some music you like. You don't want to tire yourself out, but the endorphins in your brain can help you to feel positive and confident.
- You will need water. Drink water the morning of your talk and then make sure you take a bottle with you. The stress can make your mouth dry up and you might feel hot. Taking a sip of water during your talk is completely acceptable.
- Don't drink too much water, though! There is nothing worse than needing the loo after the first ten minutes of an hour-long talk. Holding in the need to pee for an hour is painful and could end in embarrassment. Go to the toilet before.

## 11.6   Before We Move on…

Essentially, reducing your anxiety is about putting yourself in the best conditions possible to deliver your presentation. Feeling good about yourself is very helpful as is making sure your body is on your side. We don't expect you to use every single one of these tips, but we hope you find what works best for you. See what helps you to stay focused and alert, for a result you can be proud of.

# 12

## Perfect Your Performance
### Be the Best You Can Be

Let's think back to that original chocolate soufflé one last time. If we consider each chapter in this book as a step in the recipe, then this final chapter is the icing sugar that you would sprinkle over the top of the baked dessert moments before serving. In terms of your presentation, these final touches will add that extra layer of professionalism to make you stand out from the rest.

By now, you have prepared your presentation with a strong message and a clear plan. You have built upon them to make an introduction that grabs the attention of your audience, and a conclusion that will mean they won't forget you. You have added jokes and demos and impactful slides, and you have adjusted the technical information to the right level of knowledge for your audience. You have a great talk ready to present. All that is left are a few final hurdles which trip many people up on the home straight.

© Springer Nature Switzerland AG 2020
A. Youknovsky and J. Bowers, *SELL YOUR RESEARCH*,
https://doi.org/10.1007/978-3-030-34181-7_12

The second L in SELL stands for logistics, which are more critical than you may think. Improperly using a microphone or standing in the shadows on stage can make the world of difference to your talk. You shouldn't feel stressed out about these things but give yourself headspace to take notice of your surroundings. Your presentation will be great. But since we've come this far together, why not go the extra mile and make it the best it could possibly be. In this chapter, we will cover some logistical points of the actual day of your talk and what you can expect to come across.

## 12.1   Microphone Choice

A common piece of equipment you will use as a speaker is the microphone. Its sole purpose is to make your voice clearly audible to everyone in the room. Unlike theatre actors who are classically trained to project their voices into the audience, you will often be provided with a microphone. Sometimes, you will be given a choice about which type you wish to use. In other cases, you will have to use the one imposed by the organisers of the event. Either way, it is worth knowing the different types along with their advantages and disadvantages, to avoid any equipment-related hiccups.

The main types of microphone you will come across are:

- Gooseneck,
- Handheld,
- Headset,
- Clip-on.

You will likely have a technician present or one of the event organisers who will give you a hand with equipment like this. Nonetheless, be sure to check the level of battery and switch off all electronic devices that may create interference in the vicinity, namely your own mobile phone!

If you are being filmed, your microphone will likely be used by the film crew. They will simply plug into the sound board in the technical office to capture your voice for their film recording. In such a situation, before taking to the stage you should try speaking to the technicians. They will tell you where you can move without affecting the sound, while also staying in the light. If you use the microphone incorrectly, your video will be affected too.

## 12.1.1   Gooseneck Microphone

In setups where you speak behind a lectern, table or reading stand, there will be a gooseneck microphone attached to it. The term 'gooseneck' refers to the shape of the microphone: like the long, bendy neck of a goose. Hence, the microphone stand is flexible. You can simply twist it into the correct position so that it is pointed to your mouth, adapting to your height.

It is practical because the microphone is already set up on stage and your hands are free to move around. If needed, you can have your notes out in front of you. However, these microphones are quite restrictive, as you have to stay in the same place—you cannot take them with you if you move to another part of the stage. Also, you need to ensure that you don't turn your head (to look at your slides, for example), as the microphone will not pick up your voice.

Whenever possible, we advise to avoid presenting from behind a lectern or table. Regardless of the microphone issues, it is a much less inviting way to talk to a room full of people. It creates a physical barrier between you and your audience and prevents you from moving and using hand gestures; both of which are required to make your talk more dynamic.

## 12.1.2   Handheld Microphone

The handheld microphone is probably the most commonly used. With such a device, you are free to move around wherever you like: on the stage, into the audience, and you can even pass it to other people. Unfortunately, this does mean that you will only have one free hand. If you want to do a demo, hold a flash card or give out samples, you should prepare to juggle what you have in your hands.

Whilst rehearsing, if you don't have a microphone, which is very likely, you can hold a small bottle full of water instead. It is almost the same size and weight as a standard microphone. To hold it correctly: grip it along the middle of the handle, taking care not to shake or drop it to avoid making ungraceful sounds. Place the head of the microphone on your chin about two centimetres away from your mouth and keep it there. This not only means you have it at just right distance, but you will also keep it in exactly the right position if you turn your head. That way, there is no chance that the microphone will lose your voice.

### 12.1.3   Headset

Increasingly regularly, you will come across a headset microphone. Worn on your head, they could be wireless or connected to a battery pack. A headset is particularly useful if you are wishing to use a lot of space on the stage. It will be positioned on your head in such a way that your voice will be heard clearly so, usually, there is no need for you to adjust it. This type of microphone gives you a lot of freedom with your hands: you can move around or manipulate objects on the stage easily.

Do not forget that once your headset is fitted, it will stick to you. Every sound that comes out of your mouth will be audible to your audience, including coughing, sneezing or clearing your throat. You should also ensure that it's properly switched off when you aren't on stage to avoid being heard talking at an inappropriate moment.

### 12.1.4   Clip-on Microphone

Also called the 'Lavalier microphone', this microphone, is very similar to a headset and is probably the most discreet one you will come across. The small microphone is often fixed using a clip or magnet to your jacket or shirt, near the collar, to be close to your mouth. This will be attached to a battery pack by a wire, which is generally tucked away underneath your clothes. If you choose this type of microphone, fit it in advance and switch it on just before entering the stage. They can be a little fiddly, so get help from a technician when possible.

## 12.2   Chair(wo)Man

The chairman, host or master of ceremony is the person who pulls the event together, introducing scientists to the floor and driving the Q&A sessions. They, or someone from the organising team, will ask you well in advance of the day for a quick bio and a summary of your talk. This information will be used to announce you.

Generally, the chairperson will introduce themselves to you at the beginning of the event and go through the sequencing of the day. They may forget or not have time. Why not be proactive and present yourself so they know who you are. That way, you can then check your bio with them, and you could even propose a question to start off the Q&A session. This will be helpful for them too as, in the event that there are no audience questions, the chairperson will be responsible for getting them going.

Their role is also to keep on schedule. Don't be offended if they cut you off or ask you to wrap it up—you should have been within the time anyway! Sometimes they will even set up a gesture code to let you know when it's time to move on to the questions, or to conclude.

## 12.3 Technical Rehearsal

Depending on the type of event you are presenting at, you may take part in a 'dry run' of your talk. You will likely be invited to a technical rehearsal if you are entering a competition, speaking at a conference, receiving an award, giving a TED Talk or involved in a studio-based TV programme. In this type of run-through of an event, the general idea is to give everyone taking part a chance to familiarise themselves with the organisation of the day.

The amount of time dedicated to this trial run is generally proportional to the importance of everything running as smoothly as possible. A TV shoot in a studio will normally be rehearsed in great detail, as there are many people to coordinate. At a competition or conference, a quick run through of the movements is more likely. This includes where you will be seated, how and where you enter the stage, where you stand on the stage and where you leave. There may also be a test of the microphone, lighting and cameras.

You are not responsible for any of the technical details, so you need not worry about those things. The most important element for you is to know where you need to be and when. The rehearsal can sometimes be long and boring as a speaker, because most of the time is dedicated to the technical team playing around with lights and sound levels. Be patient. You could use the time to practise your speech by doing *Une Italienne,* a breathing exercise or meeting the other speakers.

You may also be in a situation where a technical rehearsal wasn't planned but there is time to do so. You could ask to practise or do a dry run on your own. It is very reassuring to have the opportunity to go through the motions of your talk, as it gives you the chance to familiarise yourself with the stage and any technical things that go with it. This helps to reduce last-minute stress. Why not, try running through the rehearsal with a friend or colleague, so they can let you know where you are standing?

Below is the checklist we use when running a technical rehearsal at events we organise.

### 12.3.1   Technical Rehearsal Checklist

- Is the sound, stage, timer, video projector, light or demonstration ready?
- Will you need to have a microphone fitted before you enter the stage? Who will do it?
- Will you be sitting in a seat at the front of the room, on the stage or backstage? Choose a seat with easy access to stage.
- Is there a host? If so, how will they call you to the stage or introduce you?
- Will the host be asking you any questions? Before or after your speech?
- How do you change slides: using a remote or a laptop? Should you bring your own equipment?
- When do you start? How will you know when your time is up?
- Are there questions at the end? If so, how long and where do you need to stand/sit at that moment?
- How do you leave the stage? Where do you sit after your presentation is over?

## 12.4   Lights, Camera…

Your organisers should inform you of whether your presentation will be filmed. It is more and more common practice for events and conferences these days, as video content of talks will be communicated online, during the event or after. If you don't hear anything about filming, you may ask so that you can avoid any nasty surprises. Depending on the size of the event, a camera crew of anywhere between one to ten people may be present. They should tell you where to position yourself on the stage, and they may even mark this out with tape on the ground. The host of the event should help you out with this too; people sometimes forget everything from the rehearsal under the pressure of standing on stage in front of a crowd.

You may have lights that you need to be aware of. A well-lit theatre setting will prevent you from seeing the audience clearly. Moreover, to avoid finding yourself or your face in the shadows, make sure you know where to stand. Finally, the lights may be warm, so dress accordingly.

If in doubt, don't worry about asking. Nobody is going to shoot you down if you ask the audience, "Can everyone see me OK in this light?". Try to iron this out in the rehearsal, but mistakes happen and sometimes things change unexpectedly at the last minute.

## 12.5 Final Touches

You don't have to be a fashion guru to look good at a presentation. However, making an effort on your appearance can only work in your favour. Not only will you feel much more confident, you will also look the part. Appropriate attire for talks will be defined by the context, but if you have ever been to a big international conference you will have noticed that what people wear varies. The most important thing is to be clean and comfortable.

At the very least, you should be wearing something you feel good in, especially if your talk will be filmed. You will find it much more difficult to watch back if your shirt wasn't buttoned up right or was wrinkled. Moreover, you should avoid any close patterns, motifs or flashy colours, as they don't look good on camera.

Do whatever else you need to do to feel good about yourself and your appearance. For some, that will mean getting to the hairdressers a few days before your talk, for others it may be a lick of make-up. Try not to get too carried away, it's just to help you look and feel fresh on the day. Your talk will be at its best if you feel legitimate.

## 12.6 That's It…

You are now at the end of your preparation. You should know everything you need to know to glide through your talk or presentation. Just remember that whatever happens, nobody is out to get you, and everything will go absolutely fine. Once you have gone through the motions of your preparation, you will also need to leave space for improvisation. Your audience may react in an unexpected way, throw you a question you hadn't prepared for or you may just need to bounce off their reactions.

Let yourself go. Play with your audience. Have fun. Break a leg!

# 13

## Final Pep Talk
## Feel Free to Re-read this Whenever You Need

Scientists present their work all the time.

With adequate preparation, it can be fantastic! A successful presentation is not only useful professionally, but it can also give you a great sense of pride. Keep your message in mind and remember that this is the most important thing to transfer to your audience. They should leave with that idea clearly imprinted in their heads.

Once you are up there, on the stage looking out over your spectators, take a deep breath. Give yourself a second to absorb the moment, making eye contact with some friendly faces in the room.

Then, when you are ready, and only then, you can begin …

© Springer Nature Switzerland AG 2020
A. Youknovsky and J. Bowers, *SELL YOUR RESEARCH*,
https://doi.org/10.1007/978-3-030-34181-7_13

# 14

## The Story Behind the SELL Method
### How This Three-Step Process Came to Be

As an instructor, I love seeing the progression in the people I train. I like to hope that those several hours we spend together will have a lasting effect. From this, the SELL Method was born. Thinking about it now, the concept of using three letters (well, four) in three steps seems logical and simple. But I assure you it didn't feel that way before it saw the light of day. Constructing this method took me much time and thought. First off, I sought to define the approach to take for preparing a good presentation. It required me to select specific tools, to remove others, to generate new ideas and to organise them into a coherent structure.

I grouped these tools into three easy-to-remember steps: Skeleton, Envelope and Life & Logistics. Then to benefit from a mnemonic, I associated them with visuals I found amusing. Once everything was strung together, I was able to test out my idea on the participants at our training courses. I can't describe how great it felt when they effortlessly took on board the SELL Method, regardless of their nationality or professional background!

The path of a trainer is full of meeting others and exchanging ideas, which helps us to progress in our practices. Therefore, here I want to thank the people who have influenced my own growth and whose input helped me to hone the SELL Method. This includes numerous trainers, all sharing the same passion for their profession and the transmission of knowledge, who have taught me a lot: Antoine Gautier, Francis Zentz, Annick de Chenay, Gérald Testé, and Eric Mathivet, to name but a few. I would also like to thank Stéphanie Barut for her daily support and Alessandra Ghione who helped us to enrich this book. Finally, I must mention Phillipe Korda, commercial negotiation expert, for whom I delivered a few training courses before creating Agent

© Springer Nature Switzerland AG 2020
A. Youknovsky and J. Bowers, *SELL YOUR RESEARCH*,
https://doi.org/10.1007/978-3-030-34181-7_14

Majeur. His ultra-efficient method for preparing negotiations made me want to create a specific technique for oral presentations too.

Indeed, my clients have played an essential role in Agent Majeur's development and our training courses. In particular, my first ever customers: Xavier Ormancey, Hervé Graindorge, Grégoire Nomitch and Thierry Leveugle. Their trust gave me confidence in myself and that state of mind gave me what I needed to earn the trust of other clients. So, thank you to everyone— new and old—who calls upon our services along with our trainees and our supportive online followers.

Whilst they are not directly involved in my work, other people have meant a lot to me over the years. My family first and my friends too, have helped me to find the energy needed to run a company on a daily basis, even during the tough times. I won't mention everyone here, but I just want to use a few lines to address my love for my two sons, Hillel and Joshua.

Last but not least, I of course have a special mention for James Bowers without whom this book would never have been written—and certainly not in English. I thank him for our collaboration throughout this journey and I am delighted that both our names are associated with this project, which means the world to me. His vision, his ideas and our exchanges brought more depth, coherence and humour to this body of work.

Public speaking can become a real pleasure, for both those who are speaking and those who are listening. That's what I hope for you from now on.

# Speaker References

Throughout this book we have used numerous examples with ideas we really liked, that have been taken from other people's presentations. Most of these cases are from individuals we have trained—but not all. Others have been picked from relevant talks, conferences and TV shows that we have enjoyed.

Needless to say, in training sessions ideas come from the speakers themselves, the coach or other participants. Therefore, even in the case of our trainees, we cannot take credit for them. To better fit the context in which the example was cited, some have been adapted and so may not entirely match the original speakers' intentions.

Here, we give details of the talks we have used which can be found in the public domain. Some presentations are not publicly available including internal communications or invitation-only external events. These we have modified significantly, and the speakers will not be cited. Nevertheless, we would like to take this opportunity to thank every single one of our trainees and any other speaker we reference for inspiring us!

**Chapter 4: Define Your Message**

*Case study II: The bionics professor,* Prof. Hugh Herr, MIT, TED Talk (2014)

**Chapter 5: Establish Your Plan**

*Case study V: The success story,* Dr. Christine Wai, The Chinese University of Hong Kong, #ShareYourAXAResearch (2018)

*Case study VI: The controversial topic,* Muriel Ehmig, SNCF, Journée doctorale (2019)

© Springer Nature Switzerland AG 2020
A. Youknovsky and J. Bowers, *SELL YOUR RESEARCH,*
https://doi.org/10.1007/978-3-030-34181-7

## Chapter 6: Hook Your Audience

*Joke: "I study the science of failure",* Jean Yves Piriou, Paris-Saclay University, Ma Thèse en 180 secondes (2015)

*Anecdote: "Chance plays an important role in scientific research…",* Dr. Salvatore Spicuglia, INSERM/Aix-Marseille Université, Fondation Bettencourt-Schueller Award Ceremony (2018)

*Analogy: "Until now, you have always listened with your ears and seen using your eyes…",* Elise Coulin, Institut Optique, Opticsvalley Fibre Prize for Innovation (2014)

*Question: "Let's all imagine that we are out in the sun. What happens to your body?",* Timothée Labouret, Université Paris-Saclay, Ma Thèse en 180 secondes (2016)

*Challenge: "…he asked everyone in the audience to take out their devices, unlock the screen and open their last text message conversation…",* Dr. Rand Hindi, CEO of Snips, TEDxSaclay (2018)

*Question: "What is the most common surgical procedure in the world?",* Dr. Guy Deniau, CEA, Opticsvalley Fibre Prize for Innovation (2013)

*Photo: "…she presented a photograph of two young girls, laughing…",* Dr. Eve Fabre, SUPAERO, #ShareYourAXAResearch (2018)

*Picture: "I work in economics, but my topic is actually more like the job of a dietician…",* Quentin Perrier, Paris Sciences et Lettres, Ma Thèse en 180 secondes (2016)

*Video: "…she included a video clip of two chimpanzees communicating via hand gestures…",* Dr. Cat Hobaiter, University of St. Andrews, Royal Institution Christmas Lectures (2017)

## Chapter 7: Design Your Slides

*Be visual: "He used a well-known image that looks like either a rabbit or a duck depending on who is looking at it…",* Jean-Pierre Tisserand, SNCF, Journée doctorale (2018)

## Chapter 8: Create Your (Additional) Supporting Materials

*Erasable boards: "He uses a whiteboard upon which he sketches a volcano and he adds content bit by bit….",* Prof. Jacques-Marie Bardintzeff, Université Paris Sud XI, Au Tableau, Universcience.tv (2010)

*3D metaphors:* "…*who carries out epidemiological studies in ageing, used a watch on her wrist to represent time…*", Dr. Archana Singh-Manoux, INSERM/Université de Paris, Fondation Bettencourt-Schueller Award Ceremony (2018)

*3D metaphors: "As a metaphor, he used a handheld, wooden bow and arrow…",* Dr. Eric Cascales, CNRS/Aix-Marseille Université, Fondation Bettencourt-Schueller Award Ceremony (2018)

*Videos: "This is what is happening in my intestine right now…",* Prof. Saiful Islam, University of Bath, Royal Institution Christmas Lectures (2016)

*Animations: "His development was a kite, which could be used to haul boats out on the open sea…",* Prof. Christophe Clanet, Ecole polytechnique, Research Thursdays (2015)

*Animations, "He used a small animation of the surface area of a zone in Europe where Beech trees grow…",* Dr. Gaël Raoul, Ecole polytechnique, Research Thursdays (2015)

# Bibliography

## English

Alley, Michael. 2003. *The craft of scientific presentations: Critical steps to succeed and critical errors to avoid.* New York: Springer.

Anderson, Chris. 2016. *Ted Talks: The official TED guide to public speaking.* New York: Houghton Mifflin Harcourt.

Antion, Tom. 1999. *Wake 'em up: How to use humour and other professional techniques to create alarmingly good business presentations.* 2nd Edition. Maryland: Anchor Publishing.

Boutang, Jérôme & De Lara, Michel. 2016. *The biased mind.* Switzerland: Springer.

Dean, Cornelia. 2009. *Am I making myself clear? A scientist's guide to talking to the public.* USA: Harvard University Press.

Duarte, Nancy. 2008. *Slide:ology.* California: O'Reilly Media.

Duarte, Nancy. 2010. *Resonate: Present visual stories that transform audiences.* USA: Wiley Publishing.

Few, Stephen. 2013. *Show me the numbers: Designing tables and graphs to enlighten.* Oakland: Analytics Press.

Frankel, Felice & DePace, Angela. 2012. *Visual strategies: a practical guide to graphics for scientists and engineers.* London: Yale University Press.

Heath, Chip & Heath, Dan. 2007. Made to stick. *Why some ideas survive and others die.* USA: Random House.

Hoff, Ron. 1996. *Say it in six: How to say exactly what you mean in six minutes or less.* Kansas City: Andrews and McMeel.

Humes, James C. 2000. *Speak like Churchill, stand like Lincoln.* New York: Three Rivers Press.

© Springer Nature Switzerland AG 2020
A. Youknovsky and J. Bowers, *SELL YOUR RESEARCH,*
https://doi.org/10.1007/978-3-030-34181-7

Lowe, Doug. 2007. *PowerPoint 2007 for dummies*. USA: Wiley Publishing.

Meredith Dennis. 2010. *Explaining research: How to reach key audiences to advance your work*. New York: Oxford University Press.

Miller, John. G. 2001. *QBQ: The question behind the question*. New York: G. P. Putnam's Sons.

Olivier, Richard & Janni, Nicholas. 2004. *Peak performance presentations*. London: JW Arrowsmith.

Pinker, Steven. 1994. *The language instinct*. USA: William Morrow and Company.

Reynolds, Garr. 2014. *Presentation zen design*. 2nd edition. USA: New Riders.

Reynolds, Garr. 2014. *Presentation zen*. USA: Pearson Education.

Van den Brul, Caroline. 2014. *Crackle and fizz: essential communication and pitching skills for scientists*. Singapore: Stallion Press.

Williams, Robin. 2010. *The non-designer's presentation book: Principles for effective presentation design*. California: Peachpit Press.

Wiseman, Richard. 2004. *Did you spot the gorilla? How to recognize hidden opportunities*. London: Arrow Books.

## French

Aguilar, Michaël. 2004. *Vendeur d'élite : Techniques et savoir-faire des meilleurs vendeurs*. 4th edition. Paris: Dunod.

Brunel, Henri. 1996. *La relaxation, c'est facile !* France: Editions du seuil.

Carnegie, Dale. 1990. *Comment parler en public*. Paris: Hachette.

Carnet, Didier. Charpy, Jean-Pierre & Creuzot-Garcher, Catherine. 2002. *La communication orale scientifique en anglais*. Paris: Ellipses. Csikszentmihalyi, Mihaly. 2006. *Vivre : la psychologie du bonheur*. (trans : Léandre Bouffard) France: Pocket.

De Peretti, André. Legrand, Jean-André & Boniface, Jean. 1994. *Techniques pour communiquer*. Paris: Hachette Livre.

Defays, Jean-Marc. 2003. *Principes et pratiques de la communication scientifique et technique*. Brussels: De Boeck

Destrez, Thierry. 2007. *Demain, je parle en public*. 4th edition. Paris: Dunod.

Gautier, Antoine. 2019. *J'ose parler en public*. Malakoff: Dunod.

Gely, Cyril. 2017. *Prendre la parole sans préparation*. 5th edition. France: Groupe Studyrama.

Kahneman, Daniel. 2012. *Système 1/Système 2 : Les deux vitesses de la pensée*. France: Flammarion.

Korda, Phillipe. 2015. Négocier et défendre ses marges, 2nd Edition. Paris : Dunod.

Lachaux, Jean-Phillipe. 2015. *Le cerveau funambule : Comprendre et apprivoiser son attention grâce aux neurosciences*. Paris: Odile Jacob.

Le Guern, Pascal & Lecaplain, Phillipe. 2007. *Savoir communiquer avec la presse*. Paris: Maxima.

Picard, Dominique. 1992. *De la communication à interaction : l'évolution des modèles*. Communication et langages. 93, pp 69-83

Potel, Catherine & Bruneau, Michel. 2006. *L'art de la formule expliqué aux scientifiques : Recommandations générales pour la rédaction de documents écrits et pour les présentations orales.* Toulouse: Cépaduès Editions.

Printed in the United States
By Bookmasters